TEXT BOOK
OF
PRACTICAL GEOGRAPHY

About the Author

Md. Zulfequar Ahmad Khan obtained his M.Sc (1988) and M. Phil. (1990) from Aligarh Muslim University, Aligarh and Ph. D. (1996) from Dr. B. R. Ambedkar University, Agra. He joined Jamia Millia Islamia (Central University) in 1990 as a Lecturer in Department of Geography. He is the life member of several reputed Geographical Journals. He has contributed more than dozen research papers published in national and international Geographical Journals as well as articles in reputed English newspapers. He has developed interest in the field of Remote Sensing, Physical Geography, Agriculture Geography and Cartography. He has also completed a certificate course on Remote Sensing Application to Town and Country Planning from Indian Institute of Remote Sensing (IIRS), Dehradun.

TEXT BOOK
OF
PRACTICAL GEOGRAPHY

MD. ZULFEQUAR AHMAD KHAN

CONCEPT PUBLISHING COMPANY, NEW DELHI - 110059

ISBN 81-7022-682-1 (HB)

First Published 1998

Published and Printed by

Ashok Kumar Mittal

Concept Publishing Company

A/15-16, Commercial Block, Mohan Garden,
NEW DELHI - 110059
Phone: 5554042, 5504042
Fax : 091-11-5598898

DEDICATED
TO
MY RESPECTED PARENTS

PREFACE

The book *"Text Book of Practical Geography"* is written basically for undergraduate and graduate level. The book provides a wealth of information and advancing knowledge of practical geography, which has to play an integrating role to science and technology with humanities. It is meant to open further frontiers of knowledge for the students and to give an understanding and the preparation which may help them to proceed further to acquire the present-day knowledge of concepts of geographical science. Precise and simple language has been used for better understanding. The comprehensive knowledge of fundamentals of practical contained in this book will enable the students to understand.

It has been tried to cover the subject in a way to provide students with a clear conception of the principles and practice of practical geography.

I am sure the book will be well received by the concerned students and academicians. Suggestions for further improvement of book will be gratefully acknowledged.

M.Z.A. KHAN

CONTENTS

Preface vii

List of Figures xiv

1. Scales 1

1.1 Definition *1*
1.2 Methods of Representation of Scale *1*
1.3 Scale Conversion *3*
1.4 Graphical Scale *5*
1.5 Plain Scale *6*
1.6 Comparative Scale *11*
(i) Different Units *13*
(ii) Time and Distance Scale *16*
(iii) Pace Scale *19*
(iv) Revolution Scale *23*
1.7 Diagonal Scale *23*
1.8 Vernier Scale *28*

2. Maps 34

2.1 Definition *34*
2.2 Classification of Maps *34*
2.3 Enlargement and Reduction of Maps *37*
2.4 Combination of Maps *42*
2.5 Measurement of Area and Distance *45*
(i) Square Method *47*
(ii) Planimeter *47*
(iii) Rotameter *49*
2.6 Uses of Maps *50*

3. Topographical Maps 51

3.1 Inerpretation of Topo-Sheet *51*
3.2 Maps of Survey of India *51*
3.3 Arrangement of Sheets on Maps of India *53*
3.4 Study of the Topographic Maps *53*
3.5 Studies of Topo-sheets (Sheet No. 63 M/15) *59*

4. Weather Maps 62

4.1 Definition *62*
4.2 Study of Weather Instruments *63*
4.3 The Indian Weather Map *78*
4.4 Elements of Weather *78*
4.5 Study of the Indian Daily Weather Report *81*

5. Geological Maps 85

5.1 Types of Geological Maps *85*
5.2 Criteria for Unfonformities *87*
5.3 Criteria for Folding *87*
5.4 Criteria for faults *87*
5.5 Thickness of Beds *87*
5.6 Exercises of Geological Maps and its Interpretations *92*

6. Slopes and Gradients 93

6.1 Kinds of Slope *93*
6.2 Methods of Representing Relief *97*
(i) C. K. Wentworth's Method of Average Slope *98*
(ii) G.H. Smith's Method of Slope Analysis *98*
(iii) The Raisz and Henry Method of Average Slope *99*
(iv) A.H. Robinson's Method of Slope Analysis *99*
6.3 Relief Representation *99*
(i) Pictorial Methods *99*
(ii) Mathemtical Methods *101*
(iii) Combination of both *103*
6.4 Relief Representation by Contours *103*

7. Map Projections 119

7.1 Definition *119*

7.2 Classification *120*
7.3 Construction of Map Projection *120*
(i) Natural Cylindrical Projection *121*
(ii) Simple Cylindrival Projection *123*
(iii) Cylindrical Equal Area Projection *124*
(iv) Mercator's Projection *126*
(v) Simple Conical Projection with One Standard Parallel *129*
(vi) Conical Projection with Two Standard Parallel *133*
(vii) Bonn's Projection *135*
(viii) Polyconic Projection *138*
(ix) Gnomonic Projection *141*
(x) Stereographic Projection *144*
(xi) Orthographic Projection *147*
(xii) Zenithal Equidistant Projection *148*
(xiii) Zenithal Equal-Area Projection *150*
(xiv) Gall's Projection *152*
7.4 Choice of Projection *156*

8. Surveying 159

8.1 Definition *159*
8.2 Classification *159*
(i) Chain Surveying *160*
(ii) Plane Table Survey *168*
(a) Radiation Method *173*
(b) Traversing Method *174*
(c) Intersection Method *174*
(d) Re-section Method *175*
(iii) Prismatic Compass Survey *177*
(a) Open Traverse *179*
(b) Close Traverse *180*
(iv) Levelling *184*
(a) Pattern of Levelling Instruments *184*
(b) Objects of Levelling *186*
(c) Differential Levelling *186*
(d) Profile Levelling *188*
(e) Ploting of the Profile *188*
(v) Box Sextant *190*
(vi) Indian Clinometer *192*
(vii) Abney's Level *194*
(viii) Theodolite *197*

9. Statistical Method 200

9.1 Measures of Central Tendency *200*
(i) Arithmetic Average or Mean *200*
(ii) Median *206*
(iii) Quartiles, Deciles and Percentiles *210*
(iv) Mode *218*
9.2 Measures of Dispersions *221*
(i) Range *221*
(ii) Quartile Deviation *223*
(iii) Co-efficient of Quartile Deviation *224*
(iv) Mean Deviation *227*
(v) Standard Deviation *232*
(vi) Co-efficient of Standard Deviation *238*
(vii) Correlation *238*
(viii) Co-efficient of Correlation *240*

10. Representation of Statistical Data 253

10.1 One Dimensional Diagram *253*
(i) Line Graph *253*
(ii) Poly or Multiple Graph *254*
(iii) Band Graph *257*
(iv) Bar Graph *257*
(v) Multiple Bar Graph *260*
(vi) Climograph or Climogram *260*
(vii) Hythergraph *263*
(viii) Ergograph *263*
(ix) Population Pyramid *267*
(x) Star Diagram *268*
(xi) Traffic Flow Cartogram *268*
10.2 Two Dimensional or Areal Diagram *270*
(i) Rectangle *270*
(ii) Square *271*
(iii) Proportional Circle *273*
(iv) Ring Diagram *276*
(v) Divided Circle *276*
(vi) Proportional Divided Circle *278*
10.3 Three Dimensional Diagrams *280*
(i) Cube Diagram *281*
(ii) Spherical Diagram *282*

10.4 Distributional Maps *285*
(i) Choropleth Map *285*
(ii) Isopleth Map *286*
(iii) Dot Map *287*

11. Study of Rocks and Minerals 288

11.1 Mineral *288*
11.2 Classification of Minerals *288*
11.3 Physical Properties of Minerals *289*
11.4 Study of Mineral in Hand Specimens *293*
11.5 Rocks *293*
11.6 Types of Rocks *293*
(a) Igneous Rock *293*
(b) Sedimentary Rock *295*
(c) Metamorphic Rock *296*
11.7 Study of Rock in Hand Specimen *297*
11.8 Some Common Minerals and Their Characteristics *303*

12. Elementary Areal Photography 312

12.1 Photogrammetry *312*
12.2 Photo Interpretation *312*
12.3 Types of Aerial Photographs *313*
12.4 General Equipment used in Air Photo Interpretation *314*
12.5 Scale Determination of Vertical Aerial Photograph *317*
12.6 Characteristics of Photographic Images *320*
12.7 Problem on Aerial Photography *321*

Appendices 323
Index 330

LIST OF FIGURES

1.1	Simple Map of a small region	3
1.2	Plain Scale to show kms.	8
1.3	Plain Scale to show kms.	8
1.4	Divisions/Sub-divisions of a line	10
1.5	Plain scale of metres and kilometers	10
1.6	Plain scale of yards and feet	12
1.7	Plain scale to show miles.	12
1.8	Plain scale of yards and miles	12
1.9	Comparative scale of miles and kms.	15
1.10	Comparative scale of yards and metres.	15
1.11	Comparative scale of miles and kms.	17
1.12	Time scale (minutes, kms.)	18
1.13	Time scale (Hours, kms.)	20
1.14	Time scale (Minutes, Miles)	20
1.15	Pace scale (Yards, Paces)	20
1.16	Pace scale (Metres, Paces)	24
1.17	Comparative Scale (Paces, Yards and Metres)	24
1.18	Revolution Scale (Metres, Revolution)	24
1.19	Secondary and Tertiary division of diagonal scale	25
1.20	Diagonal scale	27
1.21	Diagonal scale	27
1.22	Diagonal scale	29
1.23	Diagonal scale	29
1.24	Vernier scale	31
1.25	Positive Vernier scale	31
1.26	Direct Vernier scale	31
1.27	Retrograde or negative vernier scale	33
1.28	Retrograde or negative vernier scale	33

1.29 Double vernier scale 33
2.1 Reduction of map by square method 39
2.2 Enlargement of map by square method 41
2.3 Reduction and Enlargement by Triangle Method 42
2.4 Pantograph 43
2.5 Eidograph 44
2.6 Combination of map by square methods 46
2.7 Measurement of Area by square method 46
2.8 Plainmeter 48
2.9 Measurement of Area by Plainmeter 48
2.10 Measurement of Area by Rotameter 49
3.1 Arrangment of Topo-sheet 54
3.2 Arrangement of Topo-sheet 55
3.3 Convectional Signs 56
4.1 Slevension Screen 63
4.2 Centigrade Thermometer 64
4.3 Fehrenheit Thermometer 64
4.4 Wet Bulb and Dry Bulb Thermometer 65
4.5 Thermograph 66
4.6 Maximum & Minimum Thermometer 66
4.7 Rain Gauge 67
4.8 Hygrometer 69
4.9 Anemometer 69
4.10 Mercurial Barometer 73
4.11 Aneroid Barometer 74
4.12 Fortin's Barometer 75
4.13 Barograph 76
4.14 Convectional Signs 79
5.1 Geological Drift Map 86
5.2 Map to Show Unconfirmity 88
5.3 Folding Map 88
5.4 Faulting Map 88
5.5 Geological Map and its Section 90
5.6 Geological Map and its Section 91
6.1 Uniform Slop 94
6.2 Convex Slop 94
6.3 Concave Slop 96

6.4	Unducating Slop	96
6.5	Express of Slope by degree	97
6.6	Hachures, Hill Shading	100
6.7	Contours, Formlines, Trignometrical Stations and Spot Height	102
6.8	Relief Representation by Contours	104
6.9	Representation of Hill	104
6.10	Representation of Mountain, Ridge and Saddle	106
6.11	Representation of Plateau	107
6.12	Representation of Knol	107
6.13	Representation of Spur	109
6.14	Representation of Col/Pan	109
6.15	Representation of Escarpment	110
6.16	Representation of Cliff	110
6.17	Representation of Re-entrant	111
6.18	Representation of Cone and Crater	111
6.19	Representation of Gorge	113
6.20	Representation of V-shaped Valley	113
6.21	Representation of Waterfall and Rapid	114
6.22	Representation of Cirque, Hanging Valley, shaped Valley	114
6.23	Representation of Ria Coast	116
6.24	Representation of Fiord Coast	116
6.25	Representation of Sand Dunes	117
6.26	Superimpose Profile, Composit and Projected	117
7.1	Representation of Projection	119
7.2	Natural Cylindrical Projection	123
7.3	Simple Cylindrical Projection	125
7.4	Cylindrical Equal Area Projection	127
7.5	Mercator's Projection	130
7.6	Simple Conical Projection with one Standard Parallels	132
7.7	Conical Projection with two Standard Parallel	136
7.8	Bonn's Projection	138
7.9	Polyconic Projection	140
7.10	Gnomonic Zenithal projection	143
7.11	Stereographic projection	146
7.12	Polar orthographic projection	149

7.13	Zenithal Equidistant Projection	151
7.14	Zenithal Equal-Area Projection	153
7.15	Gall's Projection	155
8.1	Chain, Tape	163
8.2	Ranging Rod	164
8.3	Sketch of the area for Chain & Tape Survey	167
8.4	Plotting of Field book	169
8.5	Plane Tabe, Tripod, Alidade	170
8.6	Trough Compass	170
8.7	Radiation Method	173
8.8	Traversing Method	176
8.9	Intersection Method	176
8.10	Re-Section Method (Tracing Paper Method)	178
8.11	Prismatic Compass	178
8.12	Open Traverse	181
8.13	Close Traverse	185
8.14	Dumpy Level	185
8.15	Staff	187
8.16	Plotting of the Profile	189
8.17	Boxes Sextant	191
8.18	Plotting of the field work	191
8.19	Indian Clinometer	193
8.20	Plotting of field work	194
8.21	Abney's Level	196
8.22	Theodolite	198
10.1	Line Graph	256
10.2	Polygraph: Production of foodgrain	256
10.3	Band Graph	258
10.4	Simple Bar Diagram	259
10.5	Compound Bar Diagram	260
10.6	Sea	261
10.7	Hole	261
10.8	Clinometer of Calcutta	264
10.9	Clinometer of Madras	264
10.10	Hythergraph of Agra	265
10.11	Hythergraph of Allahabad	265

10.12 Ergograph 266
10.13 Age and Sex Pyramide 267
10.14 Directional Diagram 269
10.15 Star Diagram 269
10.16 Traffic Flow Diagram 270
10.17 Rectangle Diagram 272
10.18 Square Diagram 272
10.19 Proportional Circle 275
10.20 Ring Diagram 275
10.21 Divided/Circle or Rie/Wheel Diagram 278
10.22 Proportional Divided Circle 280
10.23 Cube Diagram 282
10.24 Sphere Diagram 283
10.25 Coropleth Map 284
10.26 Isopleth Map 284
10.27 Dot Map 285
10.28 Temperature map 286
12.1 Pocket Stereoscope 314
12.2 Mirror Stereoscope 315
12.3 Diagram of Height determination 316

1

SCALE

1.1 DEFINITION

Simply Scale is defined as the ratio between the map and their corresponding distance on the ground. It is usually shown as a line marked at regular intervals, the graduations representing proportionately larger distances on the ground. Sometimes it is also written as a ratio, for example 1:10000, which signifies that 1 unit of measurement on the map represents 10000 unit on the earth's surface.

The scale of the map depends upon the area which is to be mapped. The length of the scale varies according to the size of the map. The main purpose of the scale is to bring either the whole globe or a part of it on the convenient size of the paper. In other words, it is the scale which makes the map too handy.

According to need scale can be divided into two groups "Small Scale and Large Scale". Selection of scale depends on various factors. The important ones are as follows:

1. The size of area to be mapped,
2. The amount of details to be shown and
3. The size of the paper.

1.2 METHODS OF REPRESENTATION OF SCALE

There are three ways in which the scale is represented on the map.

(i) *The Simple Statement Method*

This is the most convenient method because its implication is direct.

The simplicity of this method is that there is not any type of calculation as well as drawing. The scale is simply written on the map. For example, 1cm = 10 kms. The merit of this scale is that, if the original drawing is to be reduced or enlarged, on reductions or enlargement the given statement scale would become inappropriate.

(ii) *The Graphical Method*

This scale is also known as constructive, graphical, plain, rod or line scale. In this method distance is represented graphically along a line, which is graduated uniformly. The graphical scale is constructed with the help of simple statement. For example, 1 cm = 10 kms. Converting the statement into graphic scale, a line of 1 cm is drawn which represents 10 kms. The merit of this method is that the scale is proportionately reduced and enlarged with reduction and enlargement of the map.

(iii) *The Representative Fraction Method (R.F.)*

In this method the numerator is always one and denominator is either one or more than one. But both values are in the same unit of measurement. For example, if the Representative Fraction (R.F.) is stated to be 1:1000, or 1/1000. This means that one unit on the map represents 1000 of the same unit on the ground. This unit may be an inch or a centimeter or any other foreign unit. If it is in centimeter, then 1 cm. on the map represents 1000 cms (0.01 km) on the ground. Thus it is a calculative method in which scale is represented by means of fraction. The merit of this method is that one can judge distances on a foreign map even if he is not familiar with the linear measurement of that country.

$$\text{Thus R.F.} = \frac{\text{Distance on the map}}{\text{Distance on the ground}}$$

Here, all the three types of scale should be given on a map for better understanding (Fig. 1.1).

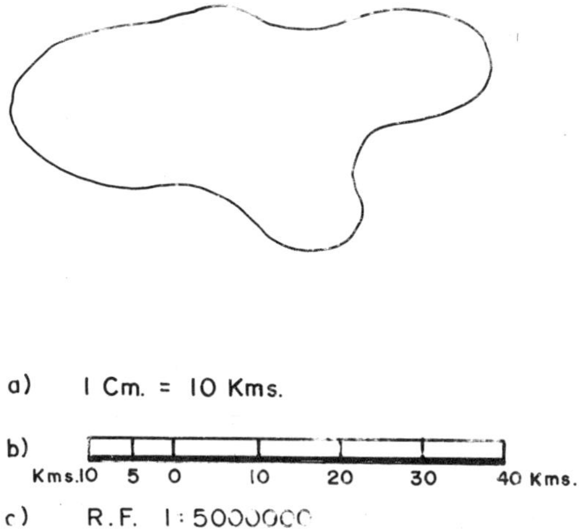

FIG. 1.1: Simple Map of a small region

1.3 SCALE CONVERSION

If the scale is known in one of the three forms, it can be expressed in the other two forms with the help of simple calculation. Georgraphically simple scale is given in two units. For example, 1 cm = 10 Kms in which both centimeters and kilometers are given since, in the R.F. numerator as well as denominator have the same unit, therefore, kilometers will be converted into centimeters to get the R.F. Few examples are given below.

Example 1: Find out the R.F. when the scale is 1" to 7 miles.

Since the numerator is an inch, the denominator should also be expressed in inches.

Now, 7 miles = 7 × 63360 inches (1 mile = 63360 inches)
= 443520

The R.F. will, therefore, be 1:443520

Example 2: Find out the R.F. when the scale is 3 inch = 1 mile.

Now, 1 mile = 63360 inches
As we know that numerator is always one therefore,
3 inches represents 63360 inches
∴ 1 inch represents 21120 inches
The R.F. will, therefore, be 1:21120

Example 3: Find out the R.F. when the scale is 1 foot = 2 miles.

Now, 2 miles = 2 × 63360 inches
= 126720
1 foot = 12 inches
1 foot represents 126720 inches
∴ 12 inches represents 126720 inches
∴ 1 inch represents 126720/12 inches
= 10560 inches
The R.F. will, therefore, be 1:10560

Example 4: Find out the R.F. when the scale is 1 cm = 2 kms.

Now 2 kms = 2 × 100000 cms (1 km = 100000 cms)
= 200000
The R.F. will, therefore, be 1:200000

Example 5: Find the R.F. when the scale is 5 cms = 10 kms.

Now 10 kms = 10 × 100000 cms
= 1000000 cms
So, 5 cms represent 1000000 cms
1 cm = 1000000/5
= 200000
The R.F. will, therefore, be 1:200000

Example 6: Find out the R.F. when scale is 0.5 cm = 2 kms.

Now 2 kms = 2 × 100000 cms
= 200000 cms
So, 0.5 cm represents 200000 cms
1 cm represents 200000/0.5 cms
= 400000 cms
The R.F. will, therefore, be 1:400000.

The R.F. may also be converted into simple statement scale. To make the statement scale into mile and kilometer, the denominator is divided by the respective units. Some examples are as follows.

Example 7: The R.F. of a map is 1:100000. Find out the scale in terms of inches and miles.

Here, 1" represents 100000 inches
1" represents 100000/63360 miles
= 1.58 miles
Hence, the require scale is 1" = 1.58 miles.

Example 8: The R.F. of a map is 1/500000. Express it in a statement of scale in the centimeter to kilometer system.

Here, 1 centimeter represents 500000 centimeters
1 cm represents 500000/100000 kilometers
= 5 kilometers
Hence, the require scale is 1cm = 5 kms.

Example 9: The R.F. of a map is 1:250000. Find out the scale in terms of centimeters to kilometers.

Here, 1 cm represents 250000 cms.
1 cm represents 250000/100000 kilometers
= 2.5 kms
The required scale is 1 cm = 2.5 kms.

Example 10: The R.F. of a map is 1:158400. Find out the scale in terms of inch to mile.

Here, 1 inch represents 158400 inches
1 inch represents 158400/63360 miles
= 2.5 miles
The required scale is 1 inch = 2.5 miles.

1.4 GRAPHICAL SCALE

As we know, in a graphical scale distances are presented along time and lengthwise, the line can be of any length. Therefore, a student is free to draw a line of any arbitrary length according to his choice.

Although, the standard length has been chosen that 6 inches in metric units and 15 cms in S.I. Unit. However, where necessary the length can have any suitable measurement.

The graphical scale can be broadly classified into four groups, they are:

1. Plain Scale
2. Comparative Scale
3. Diagonal Scale
4. Vernier Scale

1.5 PLAIN SCALE

Generally the plain scale is elaborated either with miles or with kilometers or with their fractions. In plain scale, maximum two units of distance can be shown at a time (i.e. Kilometers and Meters or Centimeters).

Illustration

Example 11: Draw a plain scale when the scale of map is 1 cm = 2 kms. Find out the R.F. and read a distance of 12 kms.

There are two steps for the solution of this problem. The first is calculation and second is construction.

Calculation

Here, 1 cm represent 2 kms
15 cms represent 2 × 15 kms
= 30 kms.
To calculate the R.F.
Now, 2 kms = 2 × 100000 cms
= 200000 cms
The R.F. will, therefore, be 1:200000

Construction

Draw a line of 15 cms and divide it into three equal parts. These divisions are known as primary divisions. To read upto a kilometer, one primary division has been sub-divided into ten equal parts, called secondary divisions. Mostly, first primary division from left hand side is subdivided for secondary divisions.Therefore, reading on the scale starts from zero (0), which is placed at the beginning of the second primary division. One primary division and two secondary divisions combinedly represents 12 kms. (Fig. 1.2).

Before the construction of a scale the following points must be kept in mind.

(i) Heading should be written in bold letters and adjusted in the middle of the scale.
(ii) R.F. should be written in between heading and the scale.
(iii) Values of primary and secondary divisions should be written exa-ctly in the direction of dividing points below the scale.
(iv) The standard width of the scale is 0.1".
(v) The Unit of the graduation or divisions should be written extreme right and left of the scale respectively.

Example 12: Construct a plain scale with the help of given R.F. 1:550000 and read a distance of 53 kilometers.

Calculation

Here 1 cm represents 550000 cms
15 cms represent 550000 × 15/100000 kms
= 8250000/100000 kms
= 82.5 kms.

But this is an odd number and so we take a round number of 80 kms and find out the length in centimeters that would represent it.

Here, 82.5 kms is represented by 15 cms
1 km is represented by 15/82.5 cms
80 kms is represented by 15 × 80/82.5 cms
= 14.5 cms.

Draw a line of 14.5 cms and divide it into eight (8) equal primary divisions each will represents 10 kms. Divide the first left hand primary division into ten equal parts each secondary divisions will represent one kilometer. For measuring 53 kms five primary divisions and three secondary divisions have to be added (Fig. 1.3).

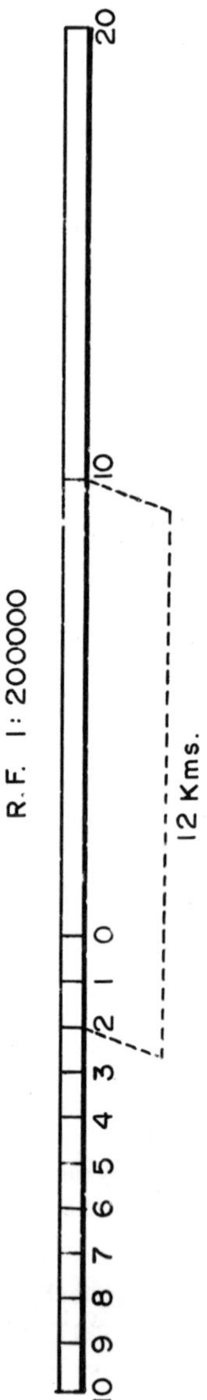

FIG. 1.2: Plain scale to show kms.

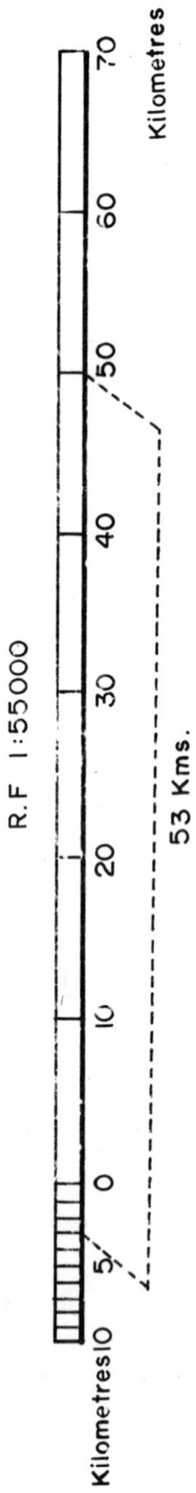

FIG. 1.3: Plain scale to show kms.

(i) Mathematical Method

14.5 cms is divided by 8
14.5/8 = 1.81 cms.

Then convert the 1.81 cms into millimeters.
Here 1.81 × 10 = 18.1 mm (1cm = 10 mm)
This 18.1 mm is represented by one primary division.
For secondary division 18.1 mm is divided by 10.
Here 18.1/10 = 1.81 mm.
This 18.1 mm is represented by one secondary division.

(ii) Geometrical Method

A line AB of 14.5 cms drawn. Another line AC is drawn at angle from the point A. AC is marked at eight equal interval. Draw a straight line BC with help of the last interval point that is C and B point. With the help of the line BC draw the parallel line one after the other from the remaining seven equal interval points ultimately these parallel line divide AB line into 8 equal parts.

By using the same method one of the primary division can be sub-divided as per need (Fig. 1.4).

Example 13: The R.F. is 1:20000, draw a scale to read kilometers and meters.

Here 1 cm represents 20000 cms
15 cms represent 15 × 20000 cms
= 300000 cms
= 300000/100000 kms
= 3 kms.

Draw a line of 15 cms and divide it into 3 equal primary divisions so that each division will represents 1 kilometer. Divide the first left hand side primary division into ten (10) equal secondary divisions, which will represent 1000 meters (Fig. 1.5).

Example 14: Draw a plain scale of yards and feets for the R.F. 1:36

Here, 1 inch represents 36 inches
6 inches represent 36 × 6 inches
= 216 inches
= 216/36 (36 inches = 1 yard)
= 6 yards.

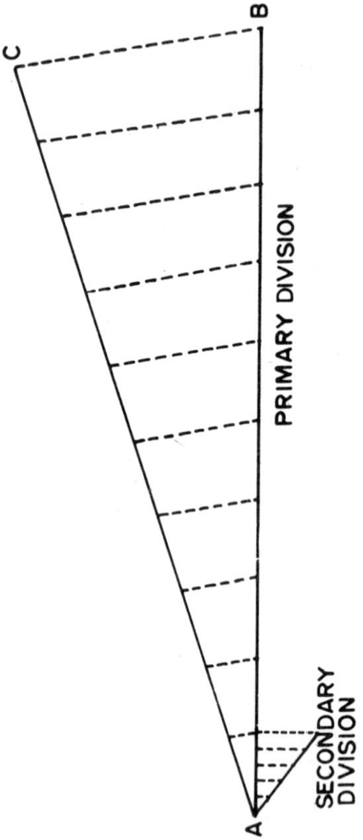

FIG. 1.4: Divisions/sub-divisions of a line

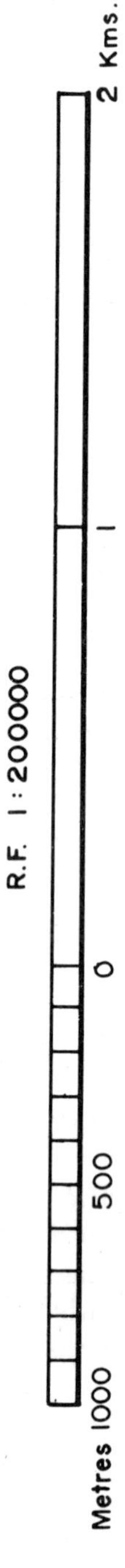

FIG. 1.5: Plain scale of meters and kms.

Draw a line of 6 inches and divide it into 6 equal parts so that each primary division will represents 1 yard. Divide the first left hand primary division into 3 equal (3 feet = 1 yard) secondary divisions (Fig. 1.6).

Example 15: The R.F. 1:105600, construct a scale to show miles

Here, 1 inch represents 105600 cms
6 inches represent 105600 × 6
=633600 inches
= 633600/63360 miles
= 10 miles.

Take a line of 6 inches which will represent the distance of 10 miles. Now 10 miles can be conveniently divided into 5 equal primary divisions of 2 miles each and the first left hand primary division into two equal secondary divisions of 1 mile each (Fig. 1.7).

Example 16: Draw a scale in miles and yards for R.F. 1:31680.

Here, 1 inch represents 31680 inches
6 inches represent 31680 × 6 inches
= 190080 inches
= 190080/63360 miles
= 3 miles.

Now take a line of 6 inches which will represent the distance of 3 miles. It can conveniently be divided into 3 equal primary divisions of 1 mile each and the first left hand primary division into 8 equal secondary divisions of 220 yards each (Fig. 1.8).

1.6 COMPARATIVE SCALE

Comparative scales popularised during the Second World War when armies of one country had to serve in other countries using a different unit of measurement. Comparative scales are also the form of plain scale in which there is a comparison of two different units of measurements (i.e. kilometers and miles, distance and time etc.). Comparative scale have also primary as well as secondary divisions.

Comparative scale is further classified on the basis of presentation.

(i) Different unit
(ii) Time scales

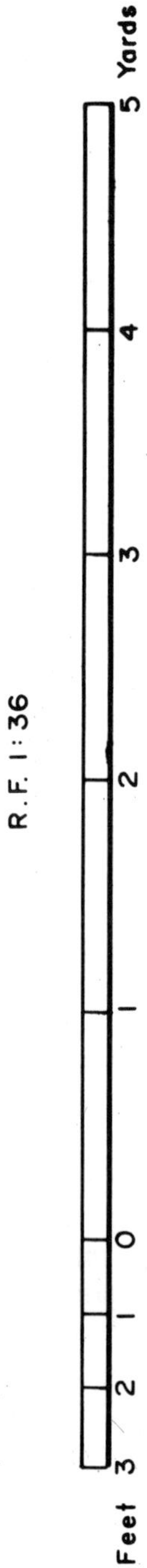

FIG. 1.6: Plain scale of yards and feet

FIG. 1.7: Plain scale of show miles

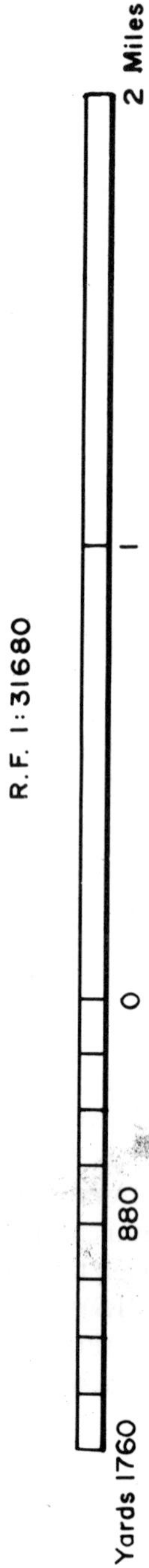

FIG. 1.8: Plain scale of yards and miles

(iii) Pace scales
(iv) Revolution scales

Some examples are as follows:

(i) Different Units

Example 17: Draw a comparative scale of miles and kilometers, when the R.F. is 1:100000.

Here, 1 cm represents 100000 cms.
15 cms. " 100000 × 15 cms.
= 1500000 cms.
= 1500000/100000 kms.
= 15 kms.
Again 1 inch represents 100000 inches
6 inches " 6 × 100000 inches
= 600000
= 600000/63360 miles
= 9.46 or 9.5 miles

But this is an odd number and so we take a round number of 9 miles and find out the length in inches that would represent it.

Here, 9.5 miles is represented by 6 inches

1 " " " 6/9.5"
9 " " " 6 × 9/9.5 inches
=5.6 inches

For representing kilometers, draw a line of 15 cms and divide it into 15 equal primary divisions, each of them will represents 1 kilometer.

For representing miles, draw a line of 5.6 inches, divide it into 9 equal primary divisions each of them will represent 1 mile. For better adjustment the second scale place just below the first one in such a manner that their zero (0) mark division coincide with each other (Fig. 1.9)

Some important things are:

(i) The units of the division of the lower scale should be written below the scale and for the upper scale it should be above the upper scale.

(ii) Zero of both the scales should coincide with each other.

Example 18: The R.F. is 1:60 draw a comparative scale to show yards and meters.

Here, 1 inch represents 60 inches

6 inches represent 6 × 60 inches

= 360 inches

= 360/36 yards

= 10 yards

Draw a plain scale measuring 6 inches. Divide it into 10 equal primary divisions so that each division will represents 1 yard. Again divide the first left hand primary division into three equal secondary divisions so that each division will represents 1 foot.
Again here, 1 cm represents 60 cms.

15 cms represents 60 × 15 cms.

=900 cms.

=900/100 meters

=9 meters.

Draw a plain scale measuring 15 cms. Divide it into 9 equal primary division so that each division will represents 1 meter. Again divide the first left hand primary division into 10 equal secondary divisions, so that each will represents 10 cms. For better adjustment, place the second scale just below the first one in such a manner that their zero (0) mark coincide with each other (Fig. 1.10).

Example 19: The scale of a map is 1" = 2 mile construct a comparative scale of miles and kilometers.

Here 1 inch = 2 miles × 63360 inches =126720 inches.

The R.F. will, therefore, be 1:126720

Here, 1 inch represents 2 miles

6 inch represent 2 × 6 miles

=12 miles

Take a line of 6 inches and divide it into 6 equal primary divisions, so that each division will represents 2 miles. Divide the first left hand

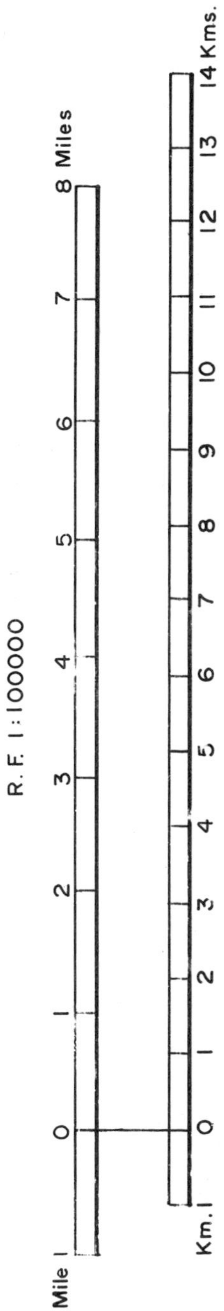

FIG. 1.9: Comparative scale of miles and kms.

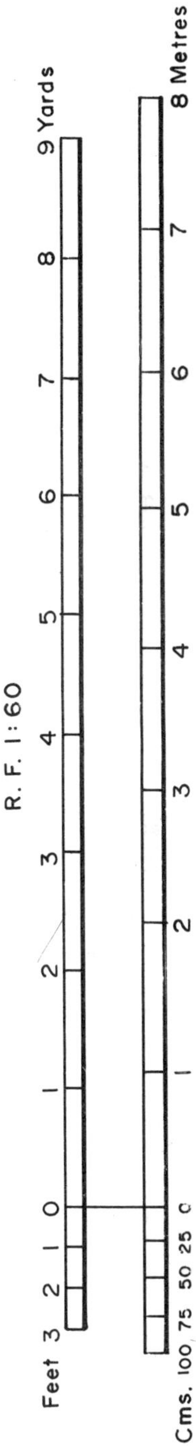

FIG. 1.10: Comparative scale of yards and meters

primary division into two equal secondary divisions. Each secondary division will represents 1 mile.

Again,

Here, 1 cm represents 126720 cms.

15 cms represent 126720 × 15 cms.

=1900800 cms.

=1900800/100000 kms.

=19.008 kms.

=19.01 kms.

But this is an odd number and so we take a round number of 19 kms and find out the length in centimeters that would represent it.

Here, 19.01 kms is represented by 15 cms.

1 km is represented by 15/19.01 cms.

20 " " " 15 × 20/19.01 cms.

=15.78 cms.

=15.8 cms.

Take a line of 15.80 cms and divide it into 10 equal primary divisions. So that each division will represents 2 kms. Divide the first left hand primary division into two equal secondary divisions. Each secondary division will represents 1 km. For better adjustment place the second scale just below the first one in such a manner that their zero (0) mark coincide each other (Fig. 1.11).

(ii) Time and Distance scale (Time Scale)

Time scale is designed in the way that it gives a time and space relationship in connection with some moving body. In other way time scale represent the relationship between time and distance. This type of scale denotes the distance travelled by a particular agent during some specific period of time. Hence, two things must be known before drawing a time scale.

1. The scale of the map of the particular area where a body moves, and.
2. The rate of movement or speed for drawing a time scale.

The following points should be taken into account as extra consideration.

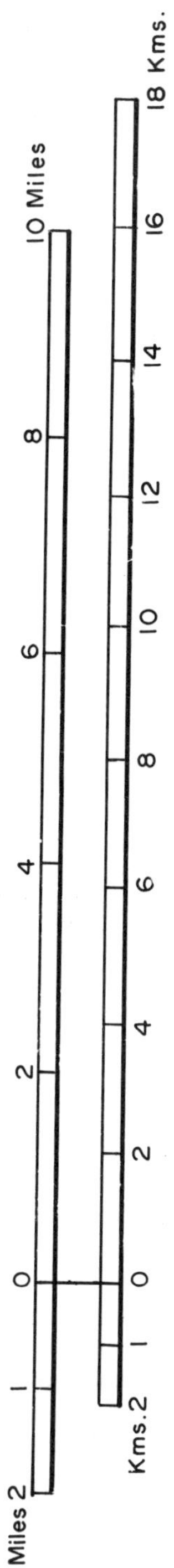

FIG. 1.11: Comparative scale of miles and kms.

FIG. 1.12: Time scale (minutes, kms)

(i) Choose such a round figure for scale which is divisible by the time unit speed because the scale should show distance and its corresponding time.

(ii) Divisions of the scale depends on;

$$\frac{\text{Round figure}}{\text{Speed}}$$

Example 20: Draw a time scale for a boy marching at 12 kilometers per hour for a map having the scale 1 centimeter to 1 kilometer.

Here, 1 kilometer is represented by 1 centimeter
12 kilometers are represented by 1 × 12 centimeters
=12 centimeters

This will be covered in one hour or 60 minutes.

Draw a line of 12 centimeters represents 1 hour which will also along with 12 kilometers and divide it into 12 equal primary divisions which will represent 1 kilometer as well as 5 minutes (Fig. 1.12).

Here, 1 cm represents 1 km
= 1 × 100000 cms.
=100000 cms.

The R.F. will, therefore, be 1:100000.

Example 21: Draw a time scale for a group of boys marching at 6 kilo-meters per hour for map having the R.F. 1.200000

Here, 6 kilometers marched in 1 hour
1 kilometer marched in 1/6 hour
30 kilometers marched in 1/6 × 30
=5 hours

Again 1 cm represents 200000 centimeters
15 cms represent 200000 × 15 cms
=3000000 cms
=3000000/100000
=30 kilometers

Draw a line of 15 centimeters which will represents 5 hours. Divide it into 5 equal primary division so that each division will represents 6 kilometers as well as one hour. Further subdivided the first left hand primary division into three equal secondary divisions so that each secondary division will represents 2 kilometers as well as 20 minutes (Fig. 1.13).

Example 22: A motorist is driving a car at the rate of 36 miles per hour. Draw a scale for a map having the scale 1 inch to 12 miles.

Here, 1 inch represents 12 miles
= 12 × 63360 inches
= 760320 inches
The R.F. will, therefore, be 1:760320
1 inch represents 12 miles
6 inches represent 12 × 6
= 72 miles
Here, 36 miles is drived in 1 hour
1 " " 1/36 hour
72 " " 72/36= 2 hours
= 120 minutes

Draw a line of 6 inches which will represent 2 hours along 72 miles. Divide it into 12 equal primary divisions so that each division will represents 6 miles as well as 10 minutes. Further sub-divide the first left hand primary division into 6 equal secondary divisions so that each secondary division will represents 1 mile as well as 100 seconds (10 minutes = 10 × 60 = 600 seconds) (Fig. 1.14).

(iii) Pace Scale

In the normal walk, the distance between paces is almost equal i.e. the standard military pace is equivalent to 2.5 feet or 30 inches. During a rapid reconnaissance or when there is no time to measure distances only paces are counted and a rough scale of distances can be constructed.

Example 23: If the length of pace is 30 inches. Draw a scale to show yards and pace for a R.F. of 1:21120

Here, 1 inch represents 21120 inches
6 inches " 21120 × 6 inches
=126720 inches
=126720/36 yards,
=3520 yads
Take the round figure 3000 yards
3520 yards are represented by 6 inches
1 yard is represented by 6/3520 inches
3000 " " " " 6 × 3000/3520 inches

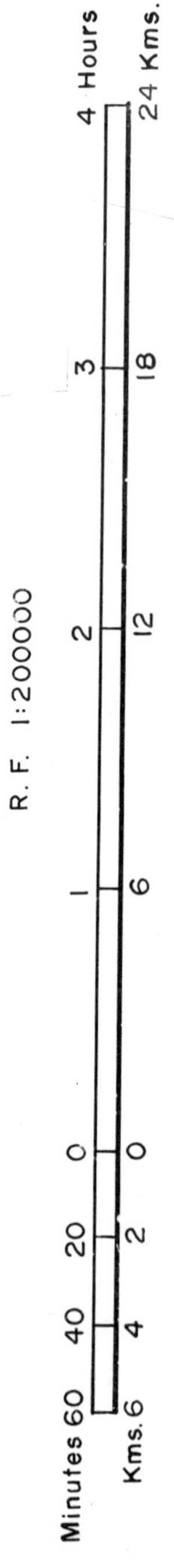

FIG. 1.13: Time scale (Hours, kms.)

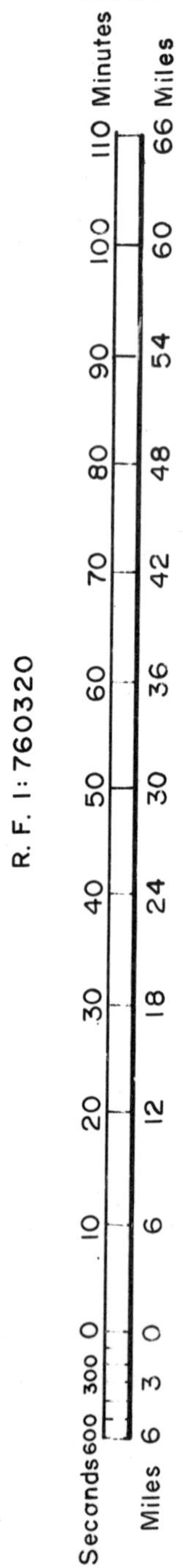

FIG. 1.14: Time scale (Minutes, kms.)

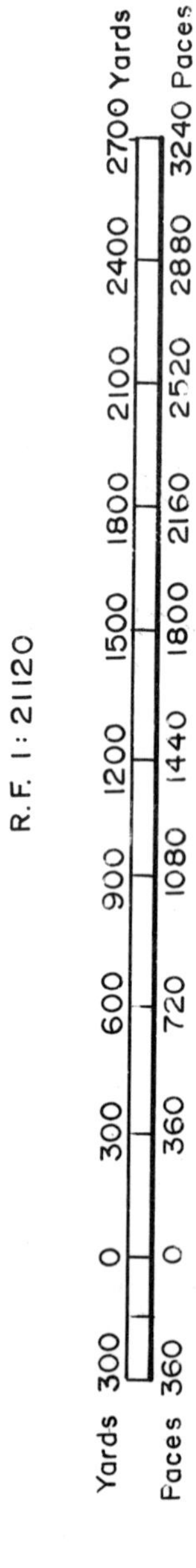

FIG. 1.15: Pace scale (yards, paces)

=5.11 inches
=5.1 inches

Here, 30 inches represent 1 pace
1 " " 1/30
3000 × 36 " " 3000 × 36/30
=3600 Paces

Draw a line of 5.1 inches which will represent 3000 yards as well as 3600 paces. Divide it into 10 equal primary divisions so that each division will represents 300 yards as well as 360 paces. Further sub-divide the first left hand primary division into 10 equal secondary division so that each division will represents 30 yards as well as 36 paces (Fig. 1.15).

Example 24: Construct a scale of paces and meters when the R.F. is 1:1000. Take 1 pace equal to 75 cms.

Here, 1 cm represents 1000 cms
15 cms " 15 × 1000 cms
=15000/100 meters
=150 meters
Here, 75 cm represent 1 pace
1 " " 1/75 paces
150 × 100 " " 150 × 100/75 paces
=15000/75 paces
=200 paces

Draw a line of 15 cms which will represents 150 meters as well as 200 paces.Divide it into 10 equal primary divisions so that each division will represents 15 meters.

Further subdivide the first left hand primary division into five equal secondary divisions so that each division will represents 3 meters as well as 4 paces (Fig. 1.16)

Example 25: Construct a comparative scale showing paces, yards and meters when the R.F. is 1:600

Suppose 30 inches = 1 pace (For yard scale)
and 75 cm = 1 pace (For meter scale)

(i) For yard scale

Here, 1 inch represents 600 inches
6 inches represent 600 × 6 inches
=3600 inches
=3600/36 yards
=100 yards

Further, 30 inches represent 1 pace
1 inch " 1/30 pace
36 × 100 " 3600/30 paces
=120 paces

Therefore, a line of 6 inches will represent 100 yards as well as 120 paces

(ii) For Meter Scale

Here, 1 cm represents 600 cms
15 cm represent 600 × 15 cms
=9000 cms
=9000/100 meters
=90 meters

Further 75 cm represent 1 pace
1 cm represents 1/75 paces
90 × 100 represent 9000/75 paces
= 120 paces

Therefore, here 15 cm line represents 90 meters as well as 120 paces.

Firstly draw a line of 6 inches which will represent 100 yards as well as 120 paces. Divide it into 10 equal primary divisions so that each division will represents 10 yards as well as 12 paces. Further subdivided the first left hand primary division into two equal secondary divisions so that each division will represents 5 yards as well as 6 paces. Secondly draw a line of 15 cms which will represents 90 meters as well as 120 paces. Divide it into 10 equal primary divisions so that each will represents 9 meters as well as 12 paces. Further subdivide the first left hand primary division so that each division will represents 3 meters as well as 4 paces. Adjust both the scale in such a manner that their zero mark division coincide with each other (Fig. 1.17).

(iv) Revolution Scale

If the country is suitable , distances can be measured rapidly with the help of any fast ground vehicle i.e. Motorcycle, bicycle etc. The circumference of the wheel is found by trial of one complete revolution along the ground. The circumference can also be found from the formula $2\pi r$ Where $\pi = 22/7$ and r = radius of the wheel.

The distance is measured by the number of revolutions of the wheel multiplied by circumference.

Example 26: A certain distance is covered by 300 revolutions of a bicycle wheel. Draw a comparative scale showing revolutions and meters on a scale 20 meters to 1 cm. The circumference of the wheel is 1 meter.

Here, 300 revolutions of wheel cover the distance of about
$300 \times 1 = 300$ meters
Further, 1 cm represents 20 meters
$= 20 \times 100$ cms
$= 2000$ cms
Therefore, the R.F. will be 1:2000
1 cm represents 20 meters
15 cms represent 20×15 meters
$= 300$ meters.

Draw a line of 15 cms which will represents 300 meters as well as 300 revolutions. Divide it into 15 equal primary divisions so that each division will represents 20 meters as well as 20 revolutions. Further subdivide the first left hand primary division into four equal secondary divisions so that each division will represents 5 meters as well as 5 revolutions (Fig. 1.18).

1.7 DIAGONAL SCALE

Diagonal scale provides much more accuracy than plain scale. Because it can represent a unit into three different multiple i.e. meters, centimeters and millimeters. In this type of scale unlike the plain scale secondary division is drawn diagonally to read the smaller fraction of the unit. Because there are tertiary divisions drawn horizontally in which each division represents upto two places of decimals. In other words, in a diagonal scale meters, centimeters and millimeters can be shown.

The drawing of divisions in diagonal scale is somewhat different.

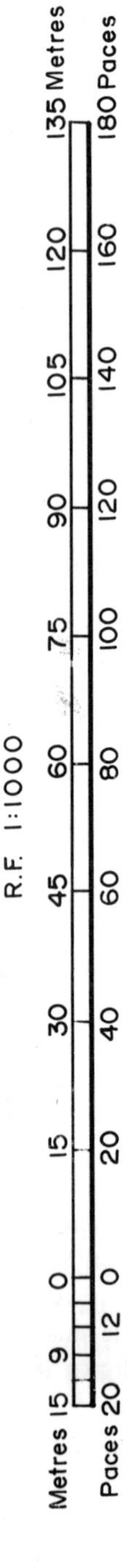

FIG. 1.16: Pace scale (metres, paces)

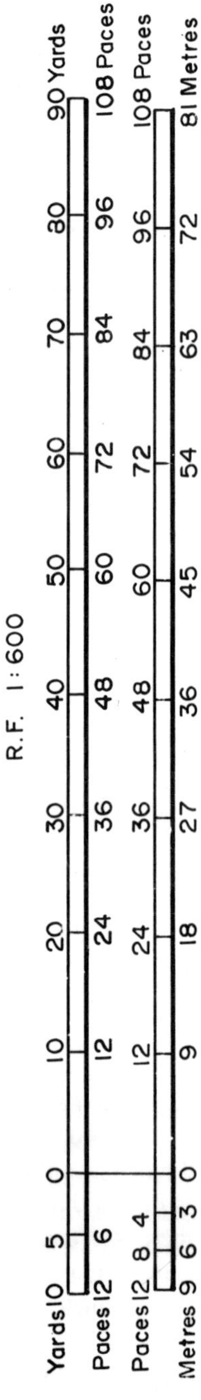

FIG. 1.17: Comparative scale (Paces, yards and metres)

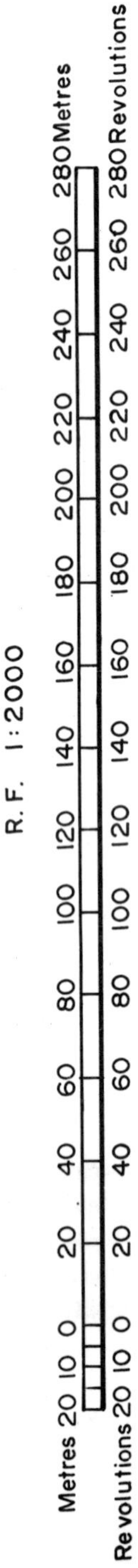

FIG. 1.18: Revolution Scale (Metres, Revolution)

Two perpendicular lines are drawn at both ends of a secondary division and these are divided into as many parts as we have already computed. Corresponding divisions on the perpendiculars are then joined by means of horizontal lines which is parallel to the secondary division. On the upper most parallel line perpendicularly point out as secondary divisions which make the small vertical rectangle. Now draw diagonal of these vertical rectangles which will give the required divisions of secondaries. Then the distance can be read correct upto the required degree of accuracy. (As shown in Fig. 1.19).

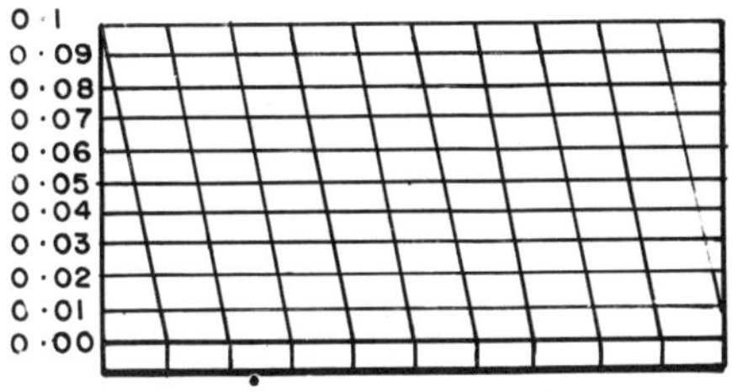

FIG. 1.19: Secondary and Tertiary division of diagonal scale

Example 27: Draw a diagonal scale for R.F. 1:20000 to read 2.47 kilo-meters.

Here, 1 cm represents 20,000 cms.

15 cms represent 20,000 × 15 cms

= 300,000 cms

= 300,000/100000 kms.

= 3 kms.

Draw a line of 15 cms which will represent 3 kms. Now 3 kms can be conveniently divided into 3 equal primary divisions of 1 km each and the first left hand division which will represents upto one place of decimal. Further draw the perpendiculars upto 1 inch of the secondary divisions and divide it into 10 equal parts. Each division will represents upto two places of decimal. Now draw the diagonal scale (as show in Fig. 1.20).

Example 28: Draw a diagonal scale for R.F. 1:40 to read meters, deci-meters and centimeters.

Here, 1 cm represents 40 cms.
15 cms represent 40 × 15 cms.
= 600 cms
= 600/100 meters
= 6 meters.

Draw a line of 15 cms, which will represent 6 meters. Divide it into 6 equal primary divisions so that each division will represents 1 meter. Now divide the first left hand primary division into 10 equal secondary divisions so that each division will represents 1 decimeter. To read upto the centimeter extend the scale perpendicular upto 1 inch. Divide it into ten equal parts. Now draw the diagonals as per the rules (Fig. 1.21)

Example 29: Draw a diagonal scale for R.F. 1:30 to read yards, feet and inches.

Here, 1 inch represents 30 inches.
6 inches represent 30 × 6 inches.
= 180 inches.
= 180/36 yards
= 5 yards

Draw a line of 6 inches, which will represent 5 yards. Divide it into 5 equal primary divisions so that each division will represents 1 yard. Now divide the first left hand primary division into 3 equal secondary divisions so that each division will represents 1 foot. To read upto the inches extend the scale perpendicular upto 1 inch. Divide it into 12 equal part so that each parts represents 1 inch, draw the diagonals as per the rules (Fig. 1.22).

Example 30: Draw a diagonal scale for R.F 1:12 to read 1.72 yards

Here, 1 inch represents 12 inches.
6 inches represent 12 × 6 inches
= 72 inches
= 72/36 yards
= 2 yards.

Draw a line of 6 inches which will represent 2 yards. Divide it into 2 equal primary divisions so that each division will represents 1

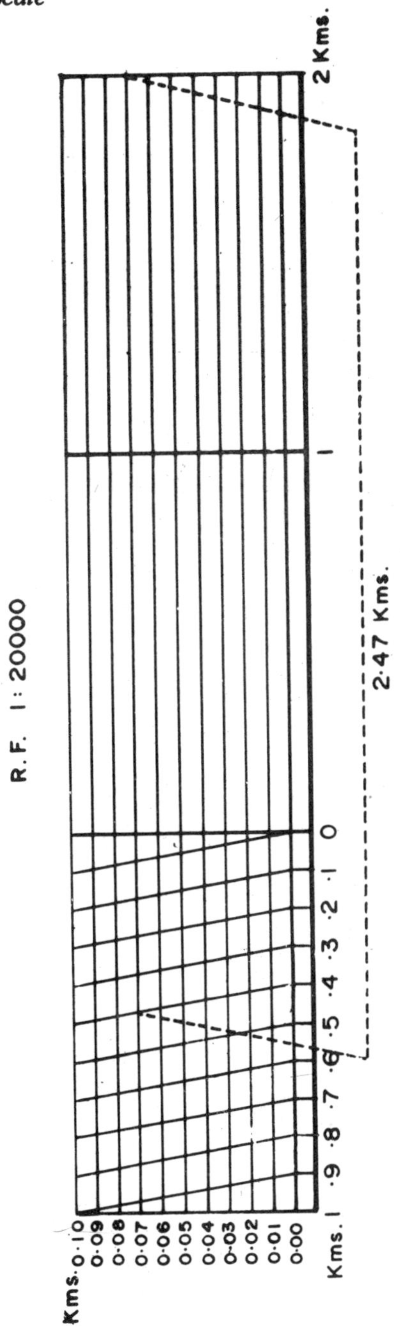

FIG. 1.20: Diagonal scale

R.F. 1 : 40

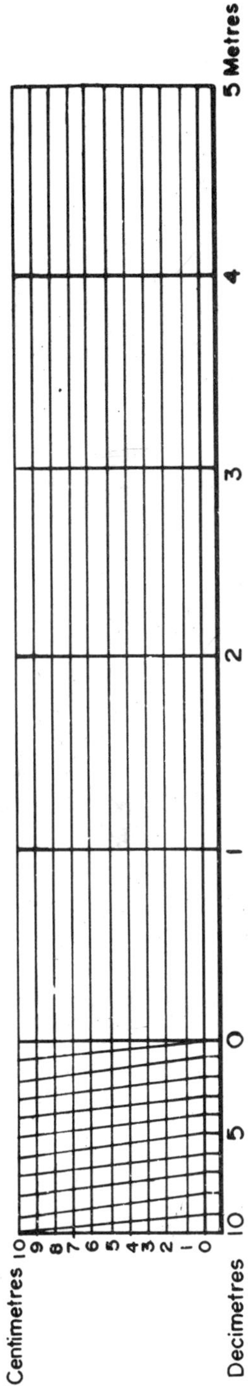

FIG. 1.21: Diagonal scale

yard. Divide the first left hand primary division into 10 equal secondary divisions so that each division will represents upto one place of decimal. Further extend the scale perpendicular upto 1 inch and divide it into 10 equal parts so that each part will represents upto two places of decimal. Draw the diagonals as per the rules (Fig. 1.23).

1.8 VERNIER SCALE

The Vernier scales enable a fraction of a division to be estimated with accuracy than plain and diagonal scale. This type of scale are used for the measurement of both linear as well as angular distances. The vernier scale consists of one small moving scale, sliding along the graduated edge of larger scale or the primary scale (Fig. 1.24) *P* for the primary scale and *V* for the vernier scale. Vernier scale is divised by Burgundian Pierre Vernier (1580 - 1637).
There are three main types of vernier scales:

(a) The Direct or Positive Vernier scale.
(b) The Retrograde or 'Negative' vernier scale.
(c) The Double vernier scale.

(a) The Direct or Positive Vernier Scale

In the Direct/Positive vernier, graduations on both the scales (Vernier as well as primary scale) run from left to right. In other words you can say in the same direction. It starts from zero which is marked in the left and increase towards right. Here each vernier is shorter by 1/nth part of the corresponding division of the primary scale.

Example 31: Construct a Direct vernier scale with a least count of 100th part of an inch for the R.F. 1:200,000.

Here, 1 cm represents 200000 cms.

15 cms represent =200000 × 15 cms

=3000000/100000 Kms.

=30 kms.

Draw a line of 15 cms which will represent 30 kms. Divide it into 15 equal primary divisions and each of them have been subdivided into 10 divisions to read upto 0.1 cm. For the constructions of vernier, 9 parts of the primary scale (which is equal to 0.9 cms) are divided

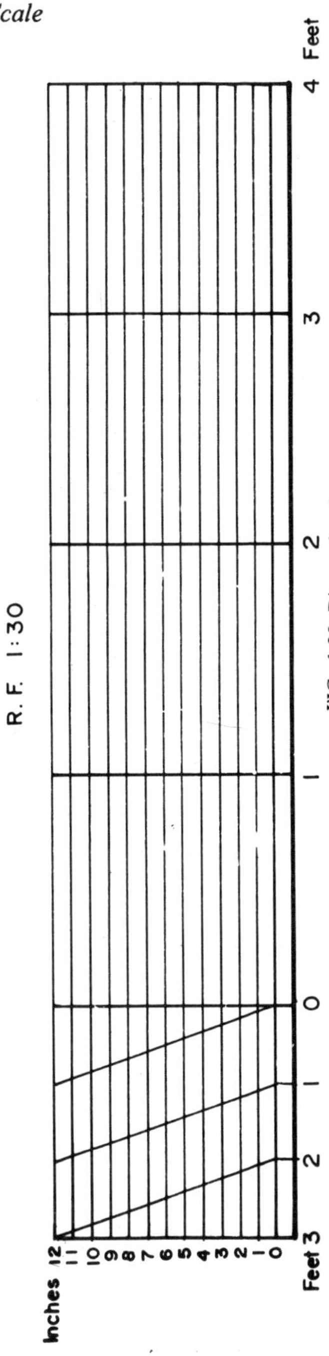

FIG. 1.22: Diagonal scale

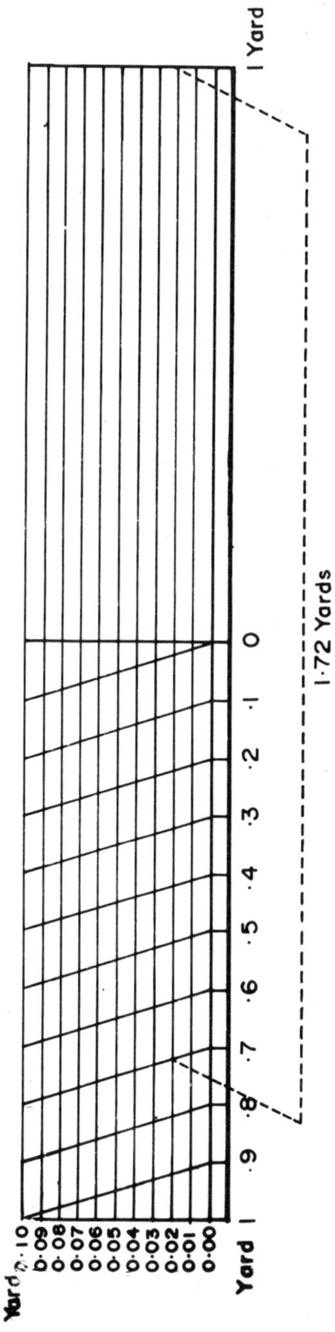

FIG. 1.23: Diagonal scale

into 10 equal divisions. Adjust the vernier scale on the primary scale in such a manner that the zero of the vernier scale coincides with the zero of the primary scale as such the first vernier division lags behind by 1/100 cms, the second 2/100 cms, the third by 3/100 cms... and the tenth by 10/100 cms or 1/10 cm or one small division of the primary scale (Fig. 1.25).

Example 32: Draw a direct vernier scale to read upto one minute and read an angle of 6° 33'.

Construction

Draw a line of 15 cms and divide it into 15 equal primary divisions so that each division will represent 1 degree. One degree which is equal to 1 cm is subdivided into two equal parts so that each subdivision will represents 30 minutes. For constructing the vernier, 29 parts of the primary scale (14.5 cms) is divided into 30 equal divisions and placed upon the primary scale in such a manner that the zero of the vernier scale coincides with the zero of the primary scale. To read 6°33', the zero of the vernier scale is moved towards right upto 6° 30'. To read 3 minutes the division of the 3 of vernier scale is slightly pushed towards right till the perpendicular adjustment to any line of the primary scale. In this way get the reading of 6° 33' (Fig. 1.26).

(b) The Retrograde or Negative Vernier Scale

In this type of vernier, the graduations on the primary scales run from left to right but the graduations of vernier run from right to left. Here also the zero is marked by an arrow pointer. The important thing is that, here each vernier division is longer by 1/*n*th part of the corresponding division of the primary scale.

Example 33: Construct a Retrograde vernier scale to read upto 1/100" and read 3.84".

Construction

Draw a line of 6 inches and divide it into 6 equal primary divisions so that each division will represents 1 inch. Further each division will be subdivided into 10 equal parts. Here vernier will be the eleven tenths (11/10) of an inch (that is equal to 1.1 inch), graduated into ten equal

FIG. 1.24: Vernier scale

R. F. 1 : 200000

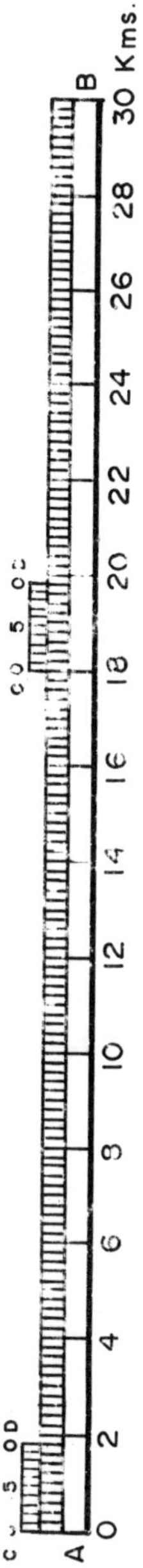

FIG. 1.25: Positive Scale

R. F. 1 : 1

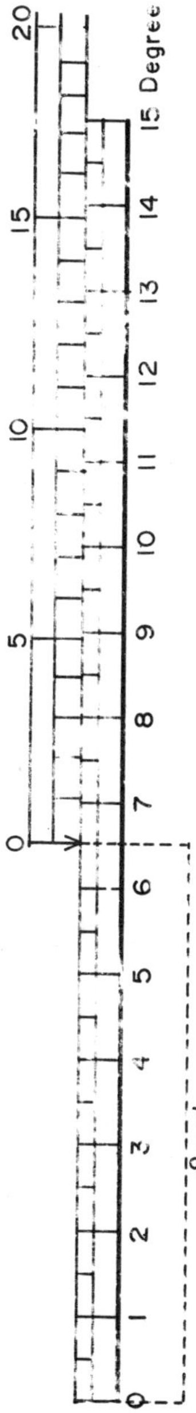

FIG. 1.26: Direct verniar scale

divisions. To read 3.84", the zero of the vernier scale, which is situated on the right side, moved towards right upto 3.84". To read .04" the division of 4 of vernier scale is adjusted perpendicular to any line of the primary scale. In this way get the reading of 3.84" (Fig. 1.27).

Example 34: Draw a Retrograde vernier scale to read upto one minute and also 10 degree 36'.

Construction

Draw a line of 15 cms and divide it into 15 equal primary divisions so that each division will represent 1 degree. One degree which is equal to 1 cm is subdivided into two equal parts so that each subdivision will represent 30 minutes. Here 31 parts of the primary scale is divided into 30 equal parts (In negative vernier, vernier 1/n is greater than primary scale).

Arrange the vernier as per the rule of negative vernier and get the reading of 10 degree 36' (Fig. 1.28).

(c) The 'Double Vernier' Scale

Two verniers are vertually used in this type of vernier scale. These two verniers are placed in the manner that constituting a single scale so that the zero and 60 stands in the mid-run and 30 in either end. This type of vernier is used for the measurement of the angular distance, one part serving for clock-wise readings and the other for anti-clock-wise readings. Its length is just double than ordinary vernier.

How to read the Double Vernier

If the vernier arrow is moved in clockwise direction, read 0 to 30, in case no mark coincide between the vernier divisions and the primary scale divisions, read it from extremely left hand side from 30 to 60. If you move the vernier arrow in anti-clockwise direction, read the vernier on the left from 0 to 30, in case no mark coincide between the vernier divisions and the primary scale divisions, read it from the extremely right hand side from 30 to 60 (Fig. 1.29). In the following figure read 72°36' or 167°24'.

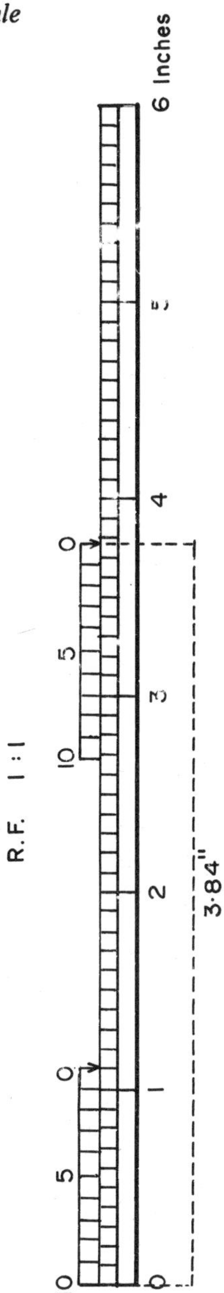

FIG. 1.27: Retrograde or negative vernier scale
R.F. 1 : 1

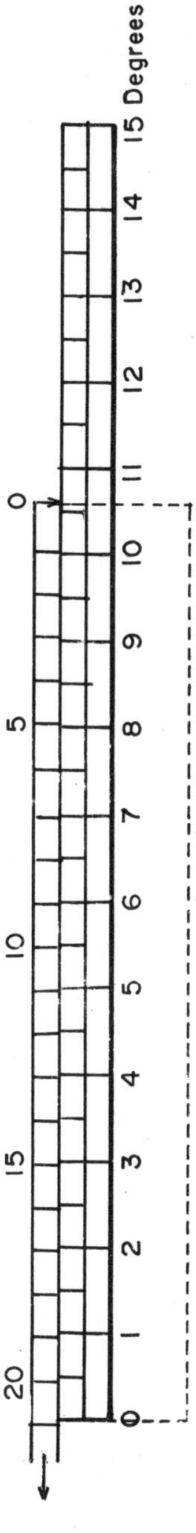

FIG. 1.28: Retrograde negative vernier scale

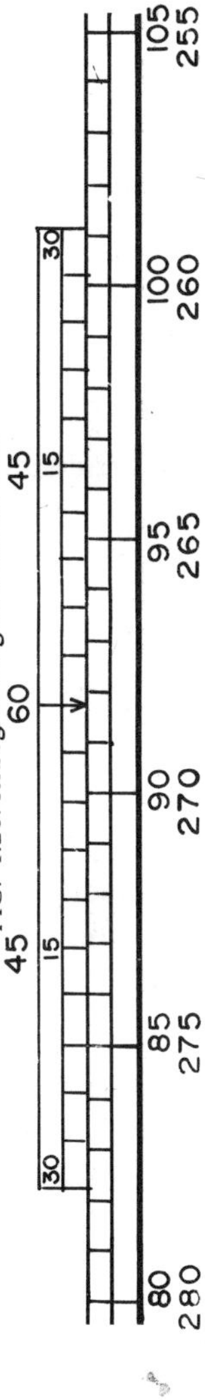

FIG. 1.29: Double vernier scale

2

MAPS

2.1 DEFINITION

A map is defined as the representation of all or part of the earth's surface or some other celestial body (i.e. Moon) on a plane surface with conventional signs. The representation is drawn to a specific scale to show distinctive aspects of the surface such as relief features, settlements, land-use, communication networks etc.

A map is drawn on plane surface so that it represents only two dimensions (length and breadth). But the earth's surface is actually curved and has three dimensions. A map is, therefore, a picture of a three dimensional curved surface on a two dimensional flat surface. The construction of maps present the problem of transferring a spherical surface to a plane surface. This problem is solved by the use of map projections. The amount of information to be represented on the map depends on:

(i) Scale
(ii) Projection
(iii) Conventional signs
(iv) Skill of the draftsman
(v) Methods of map making.

For example, a large scale as compared to the small scale, makes for big size of the map which can easily depict features of the area in much detail.

2.2 CLASSIFICATION OF MAPS

At the time of classifying maps into different categories, two essential things must be considered.

(i) Scale
(ii) Purpose

According to scale, maps may be classified as follows:

(a) Cadastral

It is used to demarcate the boundaries of fields and buildings etc. and registering the ownership of landed properties. It is especially prepared and compiled by the government agencies and is used for revenue and tax purposes. Village maps or city plan maps may be put under this category. The scale of the map varies from 16 inches to a mile or 32 inches to a mile.

(b) Topographical Maps

It is prepared on large scale to show the general surface features in detail, for example, natural landscape as well as cultural landscape. It does not show the boundaries of individual lands. It is very important tool for geographers, because, it depicts the topographical forms like relief, drainage, swamp, forest etc.

(c) Wall Maps

Wall maps are prepared for keen observation. So it is generally drawn boldly. It's scale is larger than atlas maps but smaller than topographical maps. It is generally used in a classroom.

(d) Atlas Maps

It is also known as chorographical map. It is drawn on a very small scale. It provides highly generalised information of physical, climatic and economic conditions of different regions of the earth.

According to purpose, maps are broadly classified as follows:

(1) Physical maps
(2) Cultural maps

(1) Physical Maps

Physical maps are those maps, which are especially prepared for the natural product or God gifted things such as heavenly body, soil, vegetation, relief etc. It is further sub-classified according to needs.

(a) Astronomical Map

Astronomical maps are prepared to show heavenly bodies. It may be either on large or small scale.

(b) Relief Map

It is also known as chorographic map. It is prepared to show the actual forms of the surface. It indicates the slopes, river systems, mountains, plateaus, plains etc.

(c) Geological Map

This type of map is prepared to show the types of rock, its occurrence and depositions.

(d) Climate Map

It is prepared to show the average weather condition of a long period for example fifteen years.

(e) Weather Map

It is prepared to show the average condition of temperature, pressure, wind and precipitation over a short period, which may range from a day to a season.

(f) Soil Map

It depicts the different soils of the area by different shades or colours.

(g) Vegetation Map

It is prepared to show the types as well as distribution of the various species of vegetations.

(2) Cultural Maps

This type of map is prepared to show the cultural patterns, which come into existence due to the interaction of activities of human beings with nature. Men used to modify nature according to their needs and in this way new patterns are developed. Due to differential nature, cultural maps are further sub-classified as follows:

(a) Political Map

This type of map represents the boundaries between different political units. It may be a village, block, districts, states, countries or continents.

(b) Military Map

It is prepared to show the strategic points, routes etc. for the convenience of military.

(c) Historical Map

This type of map is prepared to show the past events.

(d) Social Map

This type of map depicts, social organism (tribes and races), their languages, religions etc.

(e) Land Utilization Map

The nature and character of land-use are represented by this type of map.

2.3 ENLARGEMENT AND REDUCTION OF MAPS

Enlargement or reduction of map is chiefly done to get the appropriate size, according to needs. In other words, to get a base map on which details may be added later on to get more space for labelling further information. Reduction is done to maps of different scale to combine large scale into one.

Broadly there are three methods by the help of which map can be accurately enlarged or reduced, which are as follows:

Methods of Enlargement and Reduction of Maps

- Photographical
  - Photographic Methods
  - Photostat Methods
- Cartographical
  - Square Method
  - Similar Triangle Method
- Instrumental
  - Pantograph
  - Camera Lucida
  - Eidograph

(a) Photographic Method

This method is the superb method for enlargement and reduction. In this process, by the help of camera, negative is prepared and by the help of negative, appropriate size of photograph may be prepared.

(b) Square Method

The square method is the simplest method for enlargement and reduction of maps. It is based on the principle of adjusting distance according to scales. The map which is either enlarged or reduced is divided into any suitable network squares. Then on a sheet of paper to be reproduced on the new scale proportionately larger or smaller, find out the size of the squares required on the new scale to represent the squares on the original. The size can be found out from the proportion obtained by calculation.

New Scale/ Old scale = x

Example : To Reduce a map with a R.F. 1:10,000,000 to R.F. 1:20,000,000

New Scale/ Old scale = 1/20,000,000/1/10,000,000
= 10000000/20000000
= 1/2 inch

Draw a network of squares on the original map each with a side of 1 inch.

Now draw a network of squares each with a side of 1/2 inch, the number of squares being exactly equal to the network of original map and draw the outline square by square (Fig. 2.1).

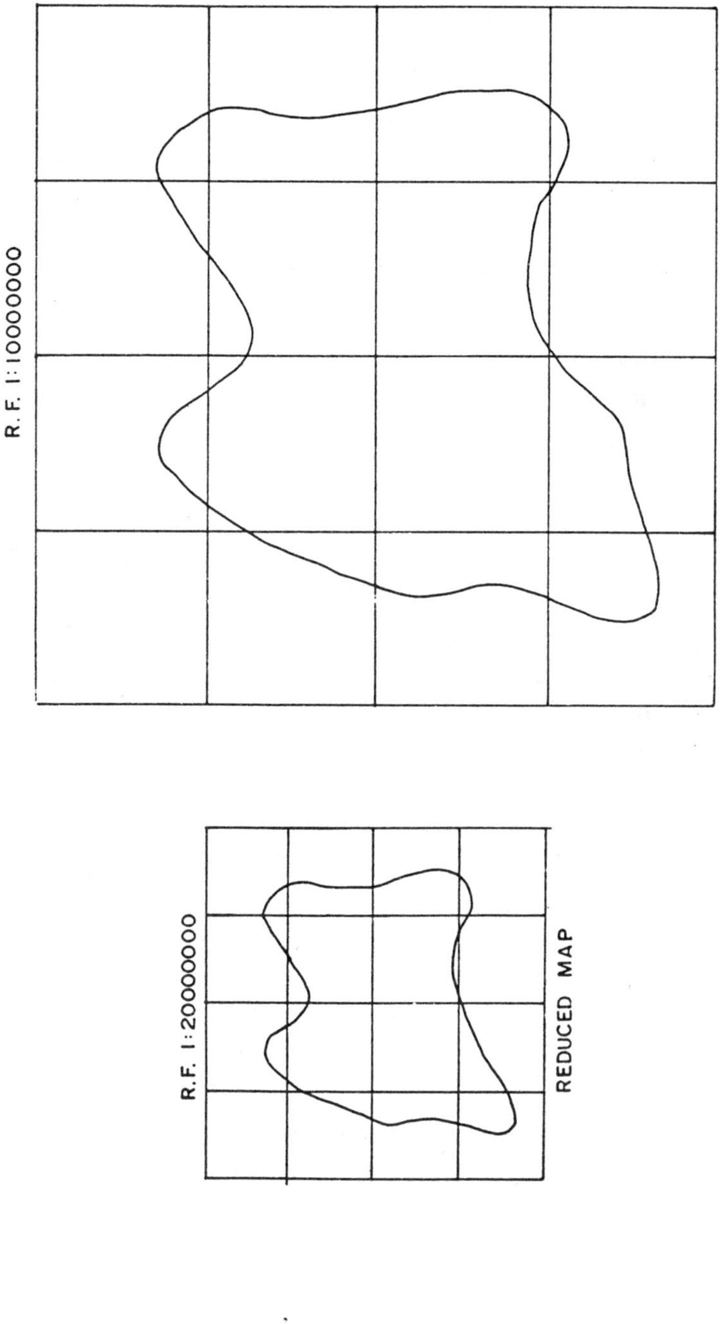

FIG. 2.1: Reduction of Map by square method

Example: Enlarge a map with a R.F. 1:18,000,000 to R.F 1:6000,000.

New Scale/Old scale

$$= \frac{1/6,000,000}{1/18,000,000} = 18,000,000/6,000,000 = 3 \text{ cms}$$

Draw the network of squares on the original map each with a side of 1 cm.

Now draw a network of squares each with a side of 3 cm, the number of squares being exactly equal to the network of original map (Fig. 2.2).

(c) Similar Triangle Method

This method is used for reducing or enlarging a narrow or linear area, such as river, railway, road, canal etc. which would otherwise present considerable difficulties. Draw a straight line *AB,* along the line of required portion. Choose any point *P* at a convenient distance from the line *AB,* the further away is the point *p,* the more accurate will be the proportion obtained. Join the curves and other specific points to *P.* If the reduction is needed five-ninth (5:9) of the original, divide *AP* into 9 equal parts and from the fifth division from *P* draw parallel line *CD* to *AB.* Join the curves and other specific points along *CD.* If enlargement is needed eleventh-ninth (11:9) of the original, divide *AP* into 9 equal parts. Extend the line *AP,* choose the point *E,* which is just double of 1/9th of the *AP.* Draw parallel line *EF* to *AB* and draw the details, like reduction (Fig. 2.3).

(d) Pantograph

In a pantograph there are two or more needles attached to a rod. When one needle is moved, the other needle moves automatically. One of the needle is moved along the outline of map, hence the other needle traces a similar outline larger or smaller as desired.

The instrument is generally used for reduction of plans, charts and maps. It can also be used for enlargement. But in case of enlargement a slight mistake in the movement of the tracer point on the original map will be several times multiplied by the enlarging needle, so that in case of enlargement, the element of error increases. Generally, a map is not enlarged more than four or fives times with a pantograph. As such, a pantograph works on the principles of the parallelogram (Fig. 2.4).

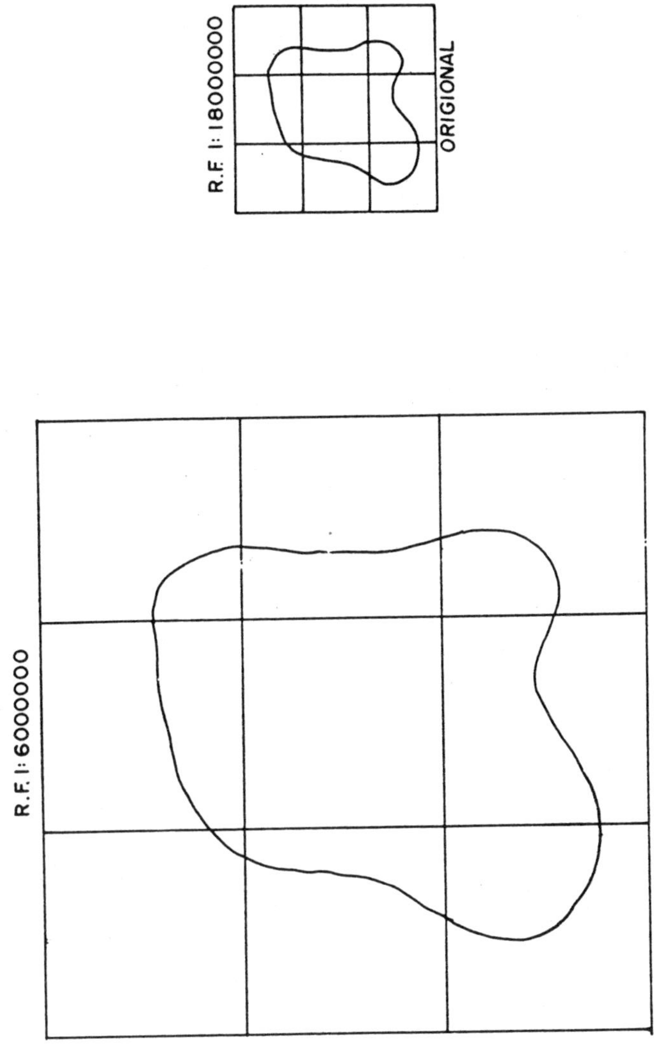

FIG. 2.2: Enlargement of Map by square method

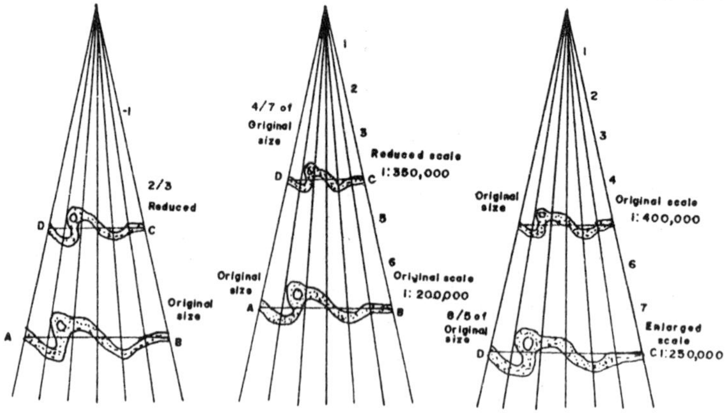

FIG. 2.3: Reduction and Enlargement by triangle method

(e) Eidograph

The eidograph having tabular form, consists of a simple frame of three graduated bars of which two bars are parallel to each other and the third one in the centre connecting both the bars, which is attached to the fulcrum fixed in a circular weight. The three bars are graduated into 100 divisions each way from the centre and the vernier on the tabular slides having the least value upto one-tenth (1/10) of a division. When the instrument is ready to work , the fulcrum, the pencil point, and the tracer must lie in straight line. Its work is based on the principles of similar triangles (Fig 2.5).

(f) Camera Lucida

It is suitable only for reduction of large maps. Its concept is based on the principles of optics and photography. The desired map either, enlarged or reduced is got by adjusting the distance of the original map and the drawing paper with respect to the prism. If the distance from the drawing paper is greater than the distance from the prism to the original map, it will lead to reduction and vice-versa.

2.4 COMBINATION OF MAPS

When the maps having different scales are used to put together, a uniform scale is constructed and combined map is drawn on that uniform scale by means of square method.

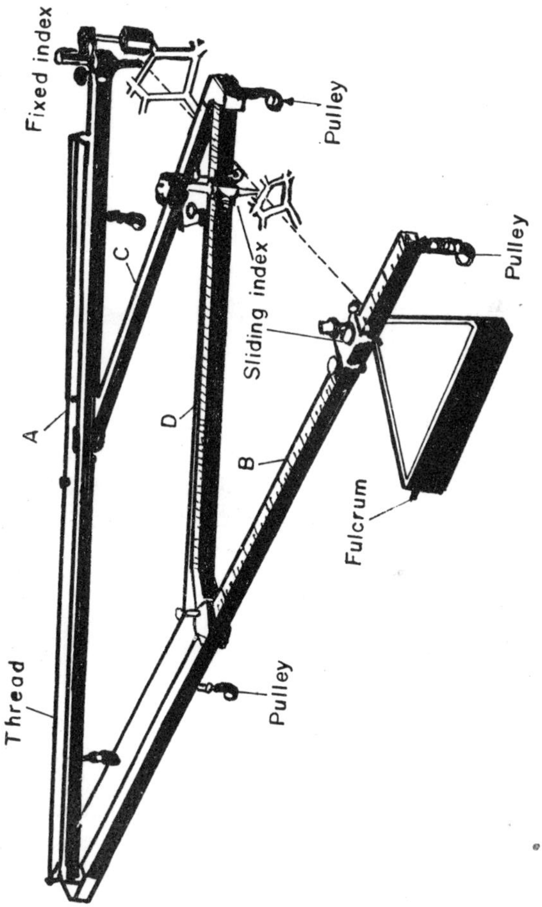

FIG. 2.4: Pantograph

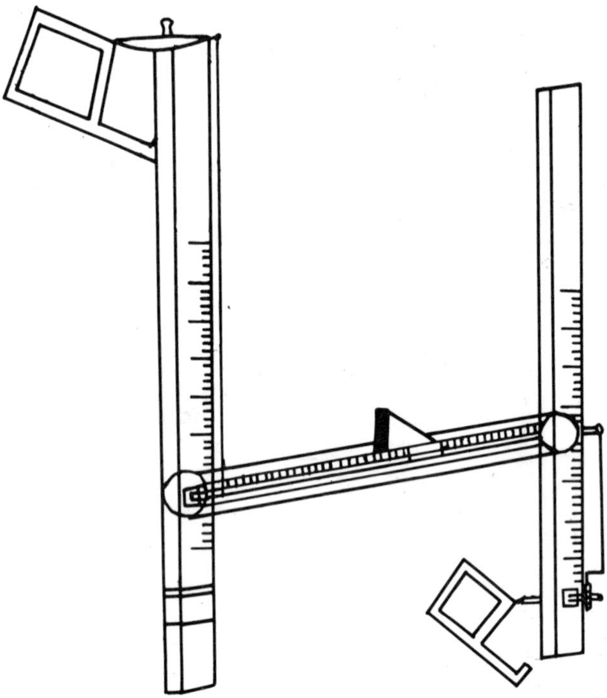

FIG. 2.5: Eidograph

Example: Combine the maps of region *A* and *B* having the scale 1:50,000 and 1:75,000 respectively on the scale of 1:100,000.

Firstly draw squares on the map of region *A* having sides 0.5 inch each.

The length of the side of the square on the region *B* is calculated by the following formula.

$$\frac{\text{Scale of Region } A \times \text{Size of Square of Region } A}{\text{Scale of Region } B}$$

$$= \frac{50{,}000 \times 0.5}{75000}$$

$$= 25/75 = 1/3 = 0.33 \text{ inch}$$

Now the side of the square for the combined map of region *A* and *B* is calculated by the following formula.

$$\frac{\text{Scale of region } A \times \text{Size of square region } A}{\text{New Scale}}$$

$$= \frac{50{,}000 \times 0.5}{100000}$$

$$= 2.5/10 = 25/100 = 0.25 \text{ inches}$$

Draw the squares on the region *A* and *B* maps with the side 0.5 inch and 0.33 inch respectively and combine them on the new network of squares drawn with the side 0.25 inch (Fig. 2.6)

2.5 MEASUREMENT OF AREA AND DISTANCE

The area of either the parts or the whole maps, charts etc. is required. Such determination can be measured through the following methods.

1. Graphical Method
2. Instrumental Method .
3. Mathematical Method.

The graphical method provides only approximate results, and is

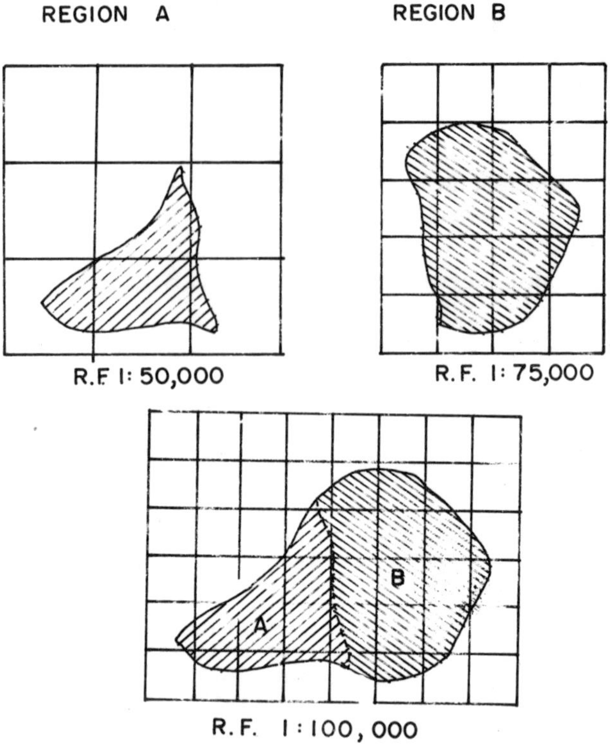

FIG. 2.6: Combination of map by square method

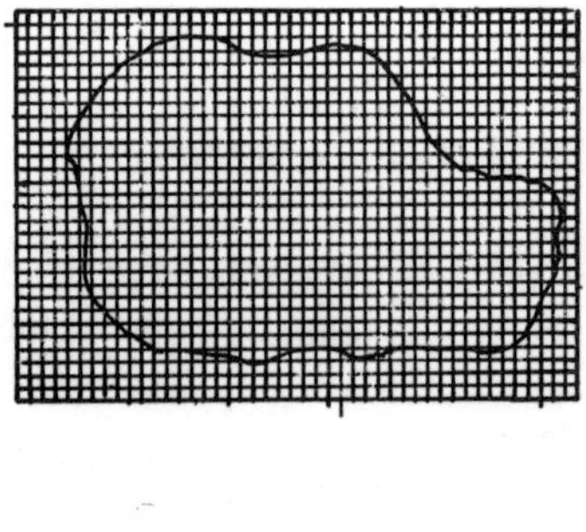

FIG. 2.7: Measurement of area by square method

time consuming too. But graphical as well as instrumental methods are easier ones. The mathematical method is the most accurate method as compared to other methods. It is applied in the case of geometrical figures only i.e. triangle, polygon etc.

(i) Square Method

Although , it is somewhat tedious, but it is a simple and easy way of determining the area of a map. To determine the area, the map is either traced on to graphpaper, or superimpose the graphpaper on the map. Firstly count the large squares, then the small one which is fully covered.

In counting the small squares, including those if more than half its area lies within the outline and those which is less than half can be neglected. Apply the scale factor to convert the area (Fig. 2.7)

Supposing a map scale of 1 cm = 2 kms. In this case 1 sq. cm = 4 sq. kms and if there be 5 complete big squares and 249 small squares within the boundary, the area of the map will be (5 × 4 sq kms) + (249 × 0.04 sq. kms) = 20 + 9.96 sq kms = 29.96 sq kms

(ii) Planimeter

It is used for the determination of areas, invented by Prof. Amster, a Swiss mathematician. It is a small, more delicate and complex instrument, which is fitted with automatic recording dials having either fixed or variable tracer arm. Tracer arm rests on drum known as measuring wheel. The graduated dial carries 10 divisions, the measuring wheel 100 divisions and vernier to read upto 1/1000 part of the measuring wheel (Fig 2.8).

Example: Find out the area of given map (Fig. 2.9) with the help of Planimeter. The scale of the map is 1:1,00,000.

Illustration

Draw a square having side 1 cm as unit area. Corresponding to scale, 1 sq. cm represents 1 sq. km. Firstly set the scale 0.000 and put the tracer point at *A*. Move the tracer in clockwise direction and bring it back to *A*. Note the initial and final readings. Repeat the process, starting from *B* and *C*, and note the initial and final readings. Calculate the differences (0.011).

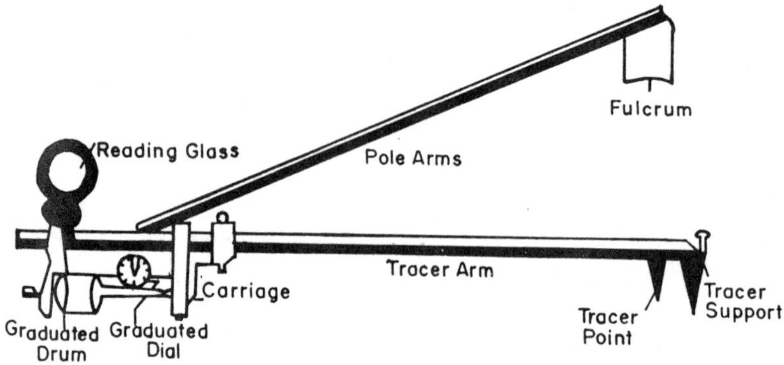

FIG. 2.8: Plainmeter

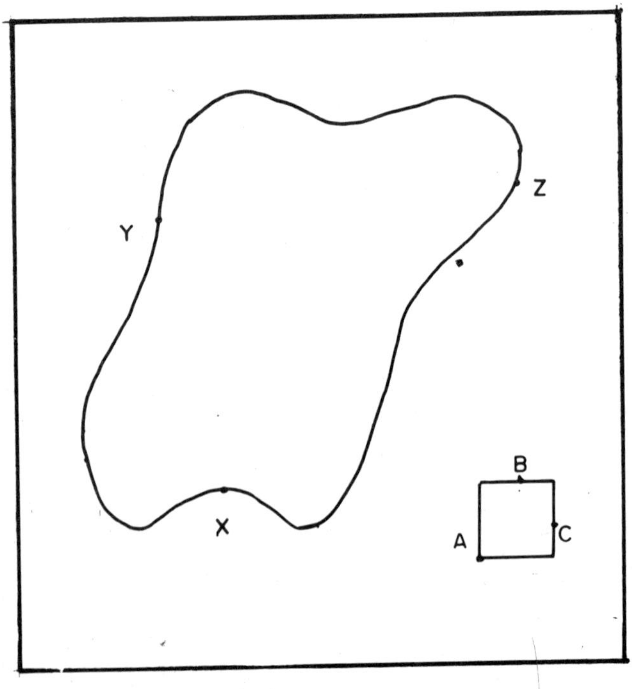

FIG. 2.9: Measurement of area by plainmeter

Now calculate the constant (K) by dividing the area of 1 sq. cm (1 sq. km) by the mean difference of unit area (0.011). Following the same steps, find out the area of given map in this case, multiply the mean difference by the constant (K).

Station	Initial Reading (IR)	Final Reading (FR)	Difference d=(FR-(IR)	Mean Difference d' (for unit area)	Mean Difference D (for whole aea)	Area of unit Sq. (A)	Constant K=A/d'	Area of Map A'=D × K
A to A	0.000	0.010	0.010					
B to B	0.010	0.021	0.011	0.011	—	4 sq km	4/0.011 =363.63	—
C to C	0.021	0.033	0.012					
X to X	0.000	0.210	0.210					
Y to Y	0.210	0.410	0.200	—	0.210	—		363.63 × 0.21 =76.36 sq km
Z to Z	0.410	0.630	0.220					

(iii) Rotameter

It is also known as opisometer and comprises of a small toothed wheel fitted with a graduated dial. It is used to measure the distance of the length of irregular line on a map to read railways, canals, rivers etc.

To determine the distance, firstly set the indicator to zero on the dial. Then tooth wheel is carefully run along the object line, the indicator moves automatically on the dial. Take, the reading from the dial in inches or centimeters (according to scale). Measure the length from both the ends and calculate the mean of the two to minimise the degree of error.

Finally the mean value is multiplied by the given scale.

Example: Find out the length of the river given in Fig. 2.10. The scale of the map is 1:200000.

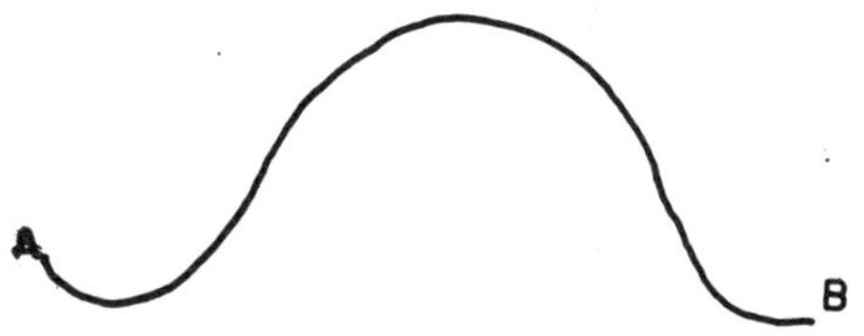

FIG. 2.10: Measurement of area by Rotameter

Illustration

Firstly the indicator of the rotameter is set at zero (0). Put the toothed wheel at point *A* and move the wheel carefully along the line *AB*. At point *B* note down the reading. Repeat the exercise from point *B* to *A*. Calculate the difference between initial and final readings. Then find out the mean value and multiply it with the scale of the map (1:200,000 or 1 cm = 2 kms).

Station	Initial reading (IR)	Final reading (FR)	Difference d=(FR-IR)	Mean difference (d')	Distance D = d' × Scale
A to B	0.00	12.00	12.00	12.00	12 × 2
B to A	12.00	24.00	12.00		=24 kms

2.6 USES OF MAPS

Maps are the essential tools of Geographers. All sciences are directly or indirectly connected with the science of the earth, So, other scientists also depend upon maps for their purposes of research. Maps are the records of various facts regarding the earth and making a direct appeal to the mind. A well illustrated map provides a lot of information in its perspective. A detail study of regional geography cannot be fulfilled without the guidance of maps. Thus, for academic purpose maps are very essential.

Maps provide a great assistance and guidance to the administrators and the travellers for their respective purposes, e.g. maps are useful to travellers and tourists in that way, they provide guidance for destination without consulting the local people. Maps are helpful in drawing the boundaries of provinces, countries etc. Thus for boundary commissions and law courts, maps are very helpful.

Without the help of navigational charts, navigator or pilot cannot move at all. Maps are even more important for the army, because during operations, maps are very helpful in indicating enemy positions.

Nowadays, in the age of planning, maps receive the prime importance, because town, regional and national planning have become practical with aims and objectives.

In the last, it may be concluded that, it is a short hand script of geographers, planners, etc.

3

TOPOGRAPHICAL MAPS

3.1 INTERPRETATION OF TOPO-SHEET

Map becomes a kind of code which cannot be fully interpreted without a complete knowledge of the conventional signs. Every symbol, sign and letter which has been adopted for representing topographic forms, convey a definite meaning. The best way to know the topography of a region is to compare the survey map of the region at the spot in the field. Although, a trained person can easily manage to know the different parts of the earth-surface in the laboratory with the help of topographical sheets.

Though map is the tool of a geographer. It is also useful and consulted by the people of other branches e.g. Geologist, traveller, military etc. But the aim of geographer is quite different. They study carefully the main features, classify them and then draw necessary conclusions from them. Thus there are two methods of approach to a map.

1. Simple approach.
2. Scientific approach.

Scientific approach is very useful to a geographer and enables to write a systematic geographical account of any area.

3.2 MAPS OF SURVEY OF INDIA

The Survey of India has been established in 1767. Since then it has published maps in a number of series.

1. Southern Asia Series

The sheets are also known as 1/M sheets or one-in-million maps. Because the scale of the map is 1:2,000,000. It covers the whole or parts of Iran, Afghanistan, Arabia, China, India, Mayanmar (Burma), Thailand and Singapore. Each sheet of this series extends over 12 degree of longitudes and 8 degree of latitudes.

2. India and Adjacent Countries Series

It includes India and adjoining countries. The scale of the map was 1:1,000,000. The sheet of this series extends over 4 degree of longitudes and 4 degree of latitudes. The map of this series has grown out of use and has been replaced by Carte International Du Monde Series. The superseded sheet has been made the base of topographic sheets of India. On that basis, the whole country is planned to be divided into 4 × 4 degree sheets, each being numbered on 51, 52, 53 etc. Some sheets are named also after the important town of the area covered by it e.g. sheet number 53 is known as Delhi Sheet.

3. Carte International Du Monde Series

Following the International agreement (London in 1909) the Survey of India has published maps of India in sheet on the International projection. In this series each sheet extends over 60 degree of longitudes and 4 degree of latitudes in which elevation is shown in meters. The scale of the map of this series is 1:1,000,000.

4. Topographical Maps of India

This type of map in sheets have been published on various scales e.g 1" maps or 1" to 1 mile; 1/2" maps or 1" to 2 miles; 1/4" maps or 1" to 4 miles.

5. Town Guide Maps

In recent years maps of important towns and tourist places have been published on comparatively larger scale that is 3" to 1 mile or 6" to 1 mile.

3.3 ARRANGEMENT OF SHEETS ON MAPS OF INDIA

The numbering of sheets in India is based on the number system of maps of India and Adjacent countries series, which bears the numbers like 1, 2, 3, ... etc. upto 136. It is known as Index Number. Each sheet is divided into 16 parts having the scale 1" to 4 miles or 1/4" to 1 mile and indexed as A, B, C, D, E, F, G, H, I, J, K, L, M, N, O and P. This quarter inch sheet is also known as degree sheet and indexed as 54A, 54B, 54C, ... 54P. The quarter inch sheets is further divided into four parts following the direction NW, NE, SE and SW and indexed as

$$54\frac{A}{NE},\ 54\frac{A}{SE}\ \ 54,\ \frac{A}{NW}\ \text{and}\ 54,\ \frac{A}{SW}$$

The scale of this is 1" to 2 miles. If the quarter inch map is divided into 16 parts, each part becomes one inch map. The numbering of the maps should be

$$54\frac{A}{1}\ \ 54\frac{A}{2}\ \ 54\frac{A}{3}\ \text{upto}\ 54\frac{A}{16}$$

For better understanding see Fig. 3.1 and (Fig. 3.2).

3.4 STUDY OF THE TOPOGRAPHIC MAPS

The study of the topo-sheets is very useful, which provides a detailed knowledge of the region. It requires a careful study. Conventional signs are very important tools and are given on the topo-sheet (Fig. 3.3). Because without conventional signs, no one can elaborate the topo-sheet. The study becomes easy, when one can follow a tentative scheme. For that, the study is classified into three broad sections, which is further divided into sub-headings.

1. Marginal Informations

A long list of informations are provided on the margins of the topo-sheet. It is necessary to know the following informations, which are listed on margins.

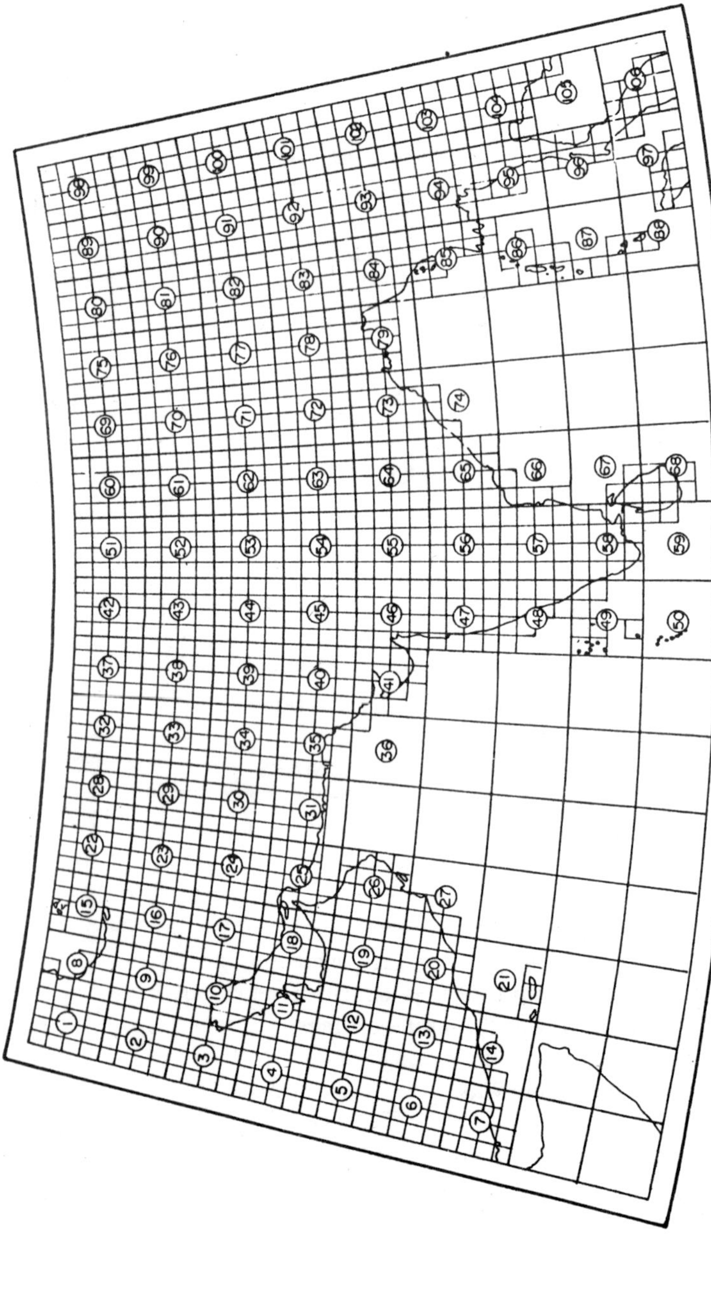

FIG. 3.1: Arrangement of Topo-sheet

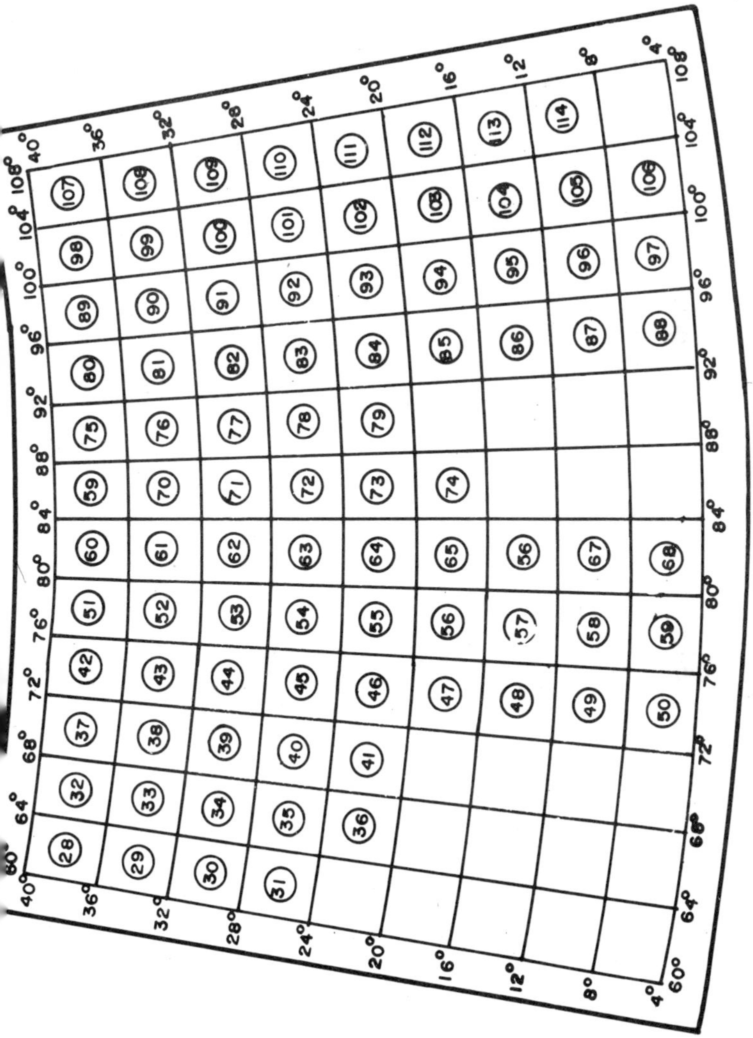

FIG. 3.2: Arrangement of Topo-sheet

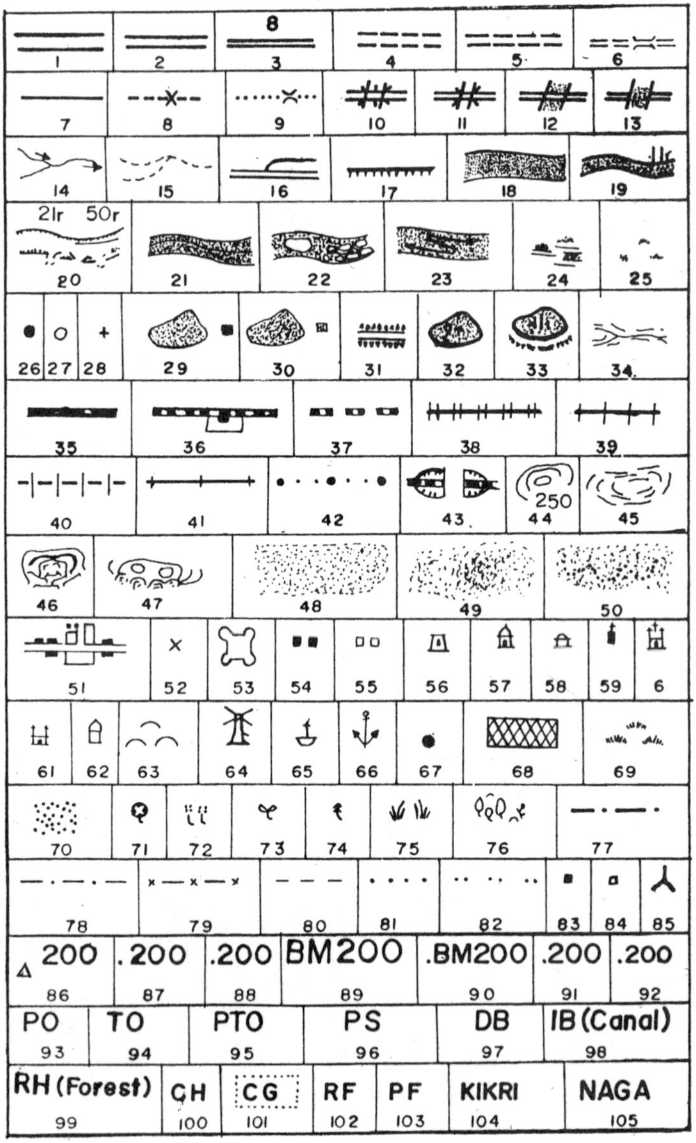

FIG. 3.3: Conventional sign.

(i) Name of the place to which the sheet belongs (e.g. state, Metropolitan City, District etc.).
(ii) Index number, year of publication and its approximate area, and
(iii) Latitudes, longitudes, scale and contour interval.

The name of the state is written on the top of the map. At the top left side the districts name as well as year of publication and the season during which the area was surveyed are noted. At the top right, the index number or sheet number as well as the direction of magnetic north and grid are given. Scale is provided in all ways (e.g. Statement; R.F. and constructive forms) at the bottom of the map.

Below the scale, the contour interval is written. There is the administrative index at the bottom right which indicates the administrative boundaries represented on the map. At the bottom left, index of convectional signs is provided. These informations are to be carefully noted.

2. Physiographic Informations

The important informations coming under this heading should be studied under the following sub-headings.

(i) Relief

First of all, examine the general nature of the relief inside the sheet with the help of the nature and behaviour of contour lines. If there is either no contour line or one or two contour lines crossing the sheet, the area is a plain. If there are several zigzag contour lines crossing the sheet, the area is a plateau. If there are several contour lines, which are not very much zigzag and height above sea level is also quite large, the area is mountainous. Other features like slope, types of coast, ridges, hills, valleys, spurs etc. can be described easily by observing the contour lines on the sheets.

(ii) Drainage

While describing the drainage, firstly describe the slope gradient by the help of contours. Because, the direction of drainage is determined by the slope. After that determine important rivers, their tributaries, the stages, type of drainage patterns, drainage density and texture.

Draw the sketch of drainage alone to know the pattern either it is a dendritic, trellis, rectangular, radial or barbed.

(iii) Vegetation

In the topo-sheet of India, natural vegetation is shown by green colour. So firstly calculate the areas covered by natural vegetation. After that, describe the type of forest, either it is protected or reserved and some other types, their distribution shown with the help of conventional signs in index.

(iv) Irrigation

Convectional signs are used to show the different kinds of means of irrigation i.e. well, tank, canals, etc. With the help of these signs, describe the system of irrigation and their subsequent effects. If there is no other kind of irrigation system, it means agriculture and vegetation depends upon rainfall only. While describing the means of irrigation also explain the cause of the kinds of irrigation, which is represented in the area e.g. well is mostly found in plateau region, because its topography is not comfortable to other sources.

3. Cultural Information

There is a good deal of cultural aspect on the topo-sheet, which includes the study of following information.

(i) Land-Use

While describing land-use, emphasis is placed on occupation i.e. cultivation, mining, forestry etc. These conclusions can be drawn on the basis of signs. For example, yellow colour is shown for the areas where agriculture is carried on and green for forest and so on. On this basis, the observer can draw conclusions.

(ii) Means of Communication

Railway lines, mineral lines, transways, ropeways, telegraph lines, roads, cart and camel tracks, ports etc. are shown on the topo-sheet with the help of conventional signs. Railway stations are marked along the track mileage and types of roads (metalled or unmetalled) as well as types of railway gauges (broad or meter gauge) are also shown. Thus, a topo-sheet provides an excellent information about the means of communication.

(iii) Settlement

While attending to settlement, firstly distinguish the types of settlement, either it is urban or rural. After that describe the important towns and cities as well as mention the population either it is dense or sparse. Thereafter, describe the site of the settlement and patterns of the settlement i.e. scattered, isolated, nucleated, clustered and linear.

3.5 STUDIES OF TOPO-SHEETS (SHEET NO. 63 M/15)

Illustration

Introduction

Survey of India published a series of one inch maps and sheet no. 63 M/15 belongs to this series. It is a graphical representation of Champaran district of Bihar, Birganj district of Nepal and Gorakhpur (Now Deoria) district of Uttar Pradesh. The area has been surveyed in 1921-22 and the scale of the map is 1 inch to 1 mile with a contour interval of 50 feet.

Relief

On account of relief features, the sheet has been divided into two parts. The north-eastern corner is hilly extending to the south along the eastern edge of the sheet and occupies north-eastern part of square *B-1,* almost whole of square *C-1* and about half of the square *C-2.* The remaining area is plain.

The hills in the north-east are low in height and are the continuations of the Siwalik ranges. The lowest contour is of 350 feet. Although, there are a large number of peaks, they have not been identified by name and only their heights have been given, which are 874', 808', 669', 828', 1011', 1538', 1660', 2625', 1270' etc. Mandar Than (height 2706 feet) is the highest peak. The ranges are not distinct but are an amalgamation of hills and valleys. Except in square *C-2,* where the hills are generally elongated with flat tops, in rest of the cases, the hills are conical with quite steep sides.

In the case of the plain, as usual no contour line is crossing it. Therefore, the features of the plain are all lower than 50 feet. The ravine lands, high banks of the Narayaini River and its deserted channel now occupied by a much smaller Rawa Nadi provide the micro relief features.

Drainage

The mountainous tract drains to the west. Great Gandak (Narayini) is the main river and its all important tributaries come from the east (Sonah Nadi, Kauna Nadi, Marha Nadi and Panchnad Nadi). The tributaries coming from the west are much smaller, hardly one mile in width and they are all seasonal, bringing water only during the rainy season. While as the tributaries coming from the east are perennial and longer ones. Elsewhere in the mountainous areas, streams are small, seasonal and significant. All rivers including Narayaini flow in the youthful stage in the mountainous area. Immediately after its descent into the plain, the Narayaini enters into maturity stage. It makes large meanders and has abandoned its channel in the south (Now occupied by Rawa nadi). In the northern part of plain, the river is in the process of straightening its course. This can be easily detected as the old course is a long curve and the new a shorter, straighter one carrying more water than the older one. Instead of formation of ox-bow lake in its full forms by the course abandoned, a small river flows there (the Rawa). There are large deposits of sand along the river course of Narayini river, which is aggrading heavily in its course. In fact, the river is braided, as it branches out into various smaller channels which meet together after a certain distance. Some tanks have turned into marshes and these can be easily identified in square *A-2* and *B-3*. Marshy areas without tanks occur mostly in square *A-3*. The channel of the Subtaman nadi in the south is itself water logged and marshy. The southern part of the Sheet is in the Tarai tract and hence the marshes.

Vegetation

The plain area in northern section of square *A*, and the entire hilly area is occupied by the forests. The belt following the course of the Narayaini Nadi is also occupied by the forsets. Forests are found in a considerable area in south, where marshes occur. There, they are known as Bharat Khand Reserved Forest and Doma Khand Reserved Forest. Forest in these areas are quite dense. In the remaining areas, the forests are absent and isolated trees occur here and there. Small grasses and isolated trees here and there cover a considerable area. Usar lands are identified where trees occur only along the river courses and where generally grass grows.

Irrigation

In this sheet, irrigations does not appear to be very prominent. The number of wells is very small and are located mostly near the settlements. Thus signifying that these wells are used more as a source of water supply than for irrigation purpose. Although there is canal also, the Tribeni canal, but it does not give out single distributary and thus does not irrigate the land covered in this sheet. Therefore, it is clear that natural supply of water which is in the form of rainfall is sufficient to fulfil the normal requirements of agriculture.

Means of Communications

The means of communication are poor and primitive in the area as shown on this sheet. Neither there are roads nor any railways. Only cart tracks are found there and these cart tracks generally do not go through the forests and cross them only where these occur in a narrow belt. Narayaini Nadi is not crossed anywhere by these tracks. These cart tracks generally connect various villages.

Settlement

The area that has been represented on this sheet shows that there exists no towns in this area. To the west of the Narayaini Nadi is Baithulia, the largest rural market in this area. Biweekly market was held on Mondays and Fridays in Baithulia. Ranghia is another small rural market worth mentioning. Almost all the villages are distributed in the cultivated area, where sweet water wells and conditions of good drainage are available. Logically these villages have sprung up near the holdings of farmers in order to facilitate them in their work of cultivation. In the west of Narayaini Nadi, the villages are larger in size than those in the east of Narayaini Nadi. These villages generally follow nucleated pattern of settlement. In square *A-3,* however, they assume a linear pattern because here they have been sheltered by well drained ground.

4

WEATHER MAPS

4.1 DEFINITION

Weather is defined as the state of atmosphere as determined by the meteorological phenomena that are occurring at any one place and time. The meteorological phenomena or weather elements include temperature, precipitation, winds, clouds, sunshine, pressure and visibility.

Temperature

Temperature denotes an index of the degree of heat of a body, substance or media such as atmosphere. It is a fundamental component of meteorology and climatology. The actual temperature of a particular place at a particular time depends on a number of factors, including latitude, season, time of day, sky cover, wind, altitude, location of sea etc. So, all temperatures are measured in a well ventilated structure known as Stevenson's Screen (Fig. 4.1) so that the thermometer remains under shade, free from the disturbances of sunshine which may raise the temperature. It is measured by thermometer. There are several types of thermometers. Several temperature scales have been devised, the celsius centigrade (C), Fahrenheit (F), Kelvin (used in scientific work) and absolute. The celsius scale is the one generally used internationally in meteorology . India has adopted the celsius scale since January 1, 1957. The weather reports give temperature in both the scales (Celsius and Fahrenheit). Generally, Celsius scale is used for upper air temperatures.

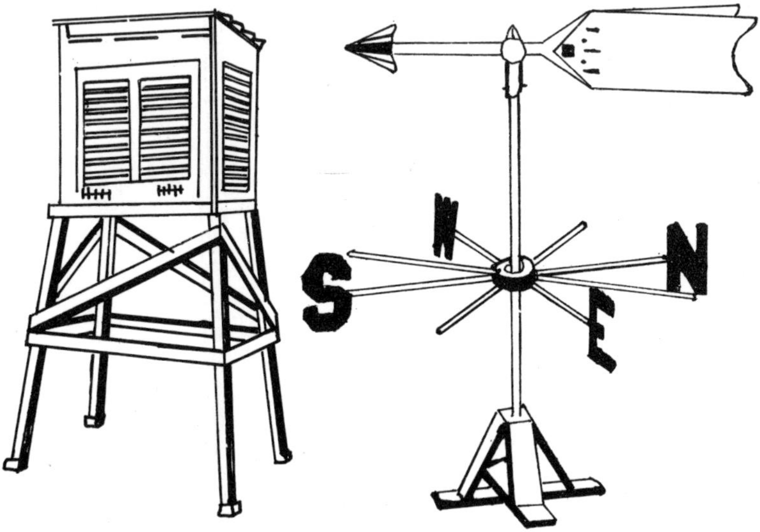

FIG. 4.1: Stevension screen and wind vane

4.2 STUDY OF WEATHER INSTRUMENTS

Centigrade Thermometer

It was named after the Swedish scientist Ander Celsius. He was the first to divide the interval between the freezing point and boiling point of water into 100 parts in 1742. In this thermometer the temperature of melting ice is taken as 0°C and the temperature of boiling water as 100° C. The interval of two points freezing and boiling is graduated into 100 equal divisions. The name was officially changed to the Celsius scale in 1948 (Fig. 4.2).

Fahrenheit Thermometer

It was named after the German physicist Gabriel Fahrenheit who introduced the scale in 1724. In this thermometer the temperature

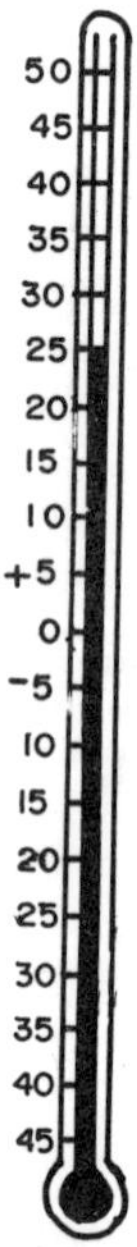

FIG. 4.2 : Centigrade Thermometer

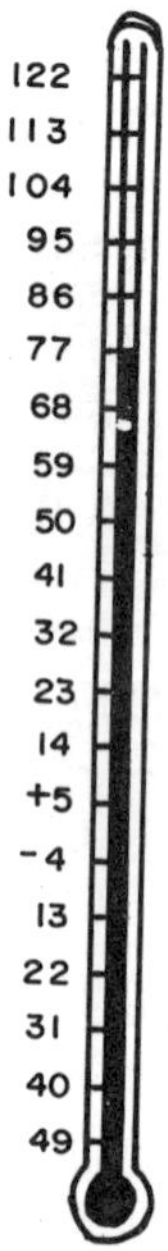

FIG. 4.3 : Fehrenheit Thermometer

32°F represents the freezing point of water and 212° F represents the boiling point of water. The interval of two points freezing and boiling as graduated into 180 equal divisions (Fig. 4.3). Thus, the ratio between centigrade and Fahrenheit thermometer is 1:1.8 (1° C = 1.8° F). For conversion of centigrade (C) to Fahrenheit (F) and vice-versa, the following formula is used:

$$C/5 = (F-32)/9$$

Wet Bulb and Dry Bulb Thermometers

It is also known as psychrometer which provides a means of measuring humidity. In the Wet-Bulb thermometer, the bulb is kept wet with a piece of a muslin, which is kept continuously moist by dipping it into a vessel of water. When the dry air touches that bulb, the temperature is lowered as compared to the Dry- bulb thermometer because of the loss of latent heat through evaporation. The drier the air, the greater will be the difference between the two temperatures shown by the bulbs. The difference between the readings in the Wet-bulb and Dry-bulb is called Wet-bulb depression. If there is no depression, it means air becomes saturated. Thus when the depression is known, humidity can be calculated easily with the help of tables (Fig. 4.4).

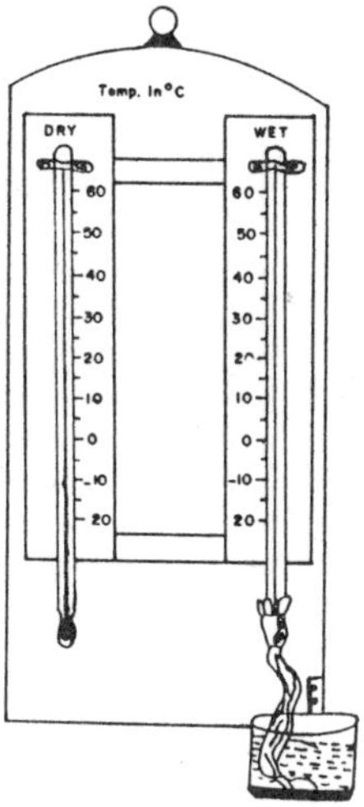

FIG. 4.4: Wet bulb and Dry bulb Thermometer

Tele Thermometer

It is used for a continuous record of temperature. With the help of this instrument the temperature outside a building can be recorded inside.

Thermograph

It is a self recording thermometer. It consists of a heat sensitive mechanism to which a pen is connected, which rests on a rotating drum. It produces a continuous trace of the changing temperature. The most common type of thermometer used is the bimetallic coil type, which expands and contracts with rising and falling temperature. The movement being transmitted through a lever to the pen (Fig. 4.5).

Maximum and Minimum Thermometers

This type of thermometer records maximum and minimum temperatures during the last 24 hours. The mean temperature is obtained by taking averages of the maximum and minimum temperature. The glass tube of the thermometer is shaped like 'U' letter of the English alphabet. The U-shaped glass tube consists of two bulbs on the top of both limbs. The graduation of the limbs are opposite to each other. Usually, the right limb records the maximum temperatures in which reading increases from bottom as against the left limb whose reading decreases from bottom. The lower part of the tube is fitted with mercury and the

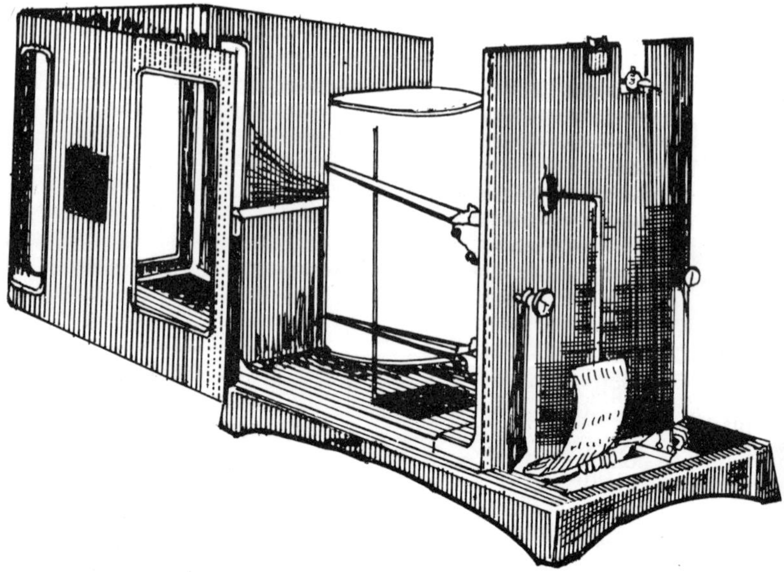

FIG. 4.5: Thermograph

remaining parts except the half portion of the right side bulb is fitted with alcohol. In each limb there is a small piece of metal (I_1 and I_2) on the mercury surface. The thermometer is set for every day by bringing down the metal pieces to the mercury level by the help of a magnet. As the temperature increases, the alcohol of the left limb starts expanding and pushing down the mercury which ultimately pushes up to the I_2 in right limb. When the temperature decreases the alcohol of the left limb contracting ultimately I_2 is pushed up. Thus, note down the minimum and maximum temperature from the lower limit of I_2 and I_1 (Fig. 4.6).

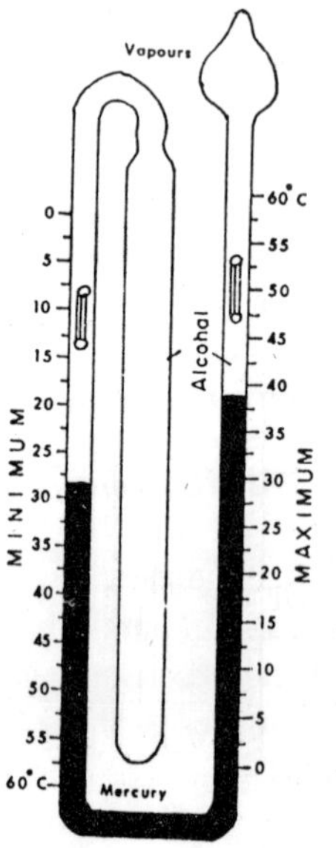

FIG. 4.6: Maximum and Minimum Thermometer

Precipitation

It is defined as the particles of water or ice that form within clouds and fall towards the earth surface. The main types are rain, drizzle, sleet, snow and hail. Any of these may evaporate before reaching the ground surface and appear as streamers below the cloud base. This phenomenon is known as virga. The amount of rainfall is measured in a rain gauge.

Rain Gauge

It is an instrument designed to measure rainfall. It consists of a funnel fitted into a collecting vessel (Fig. 4.7). The ter which falls on the funnel is collected in the specially graduated cylinder and measured. In order to obtain accurate readings, the gauge must be carefully sided, away from shelter and at a height that prevents splashing from the ground. The protective outer casing must be closely fitted in order to

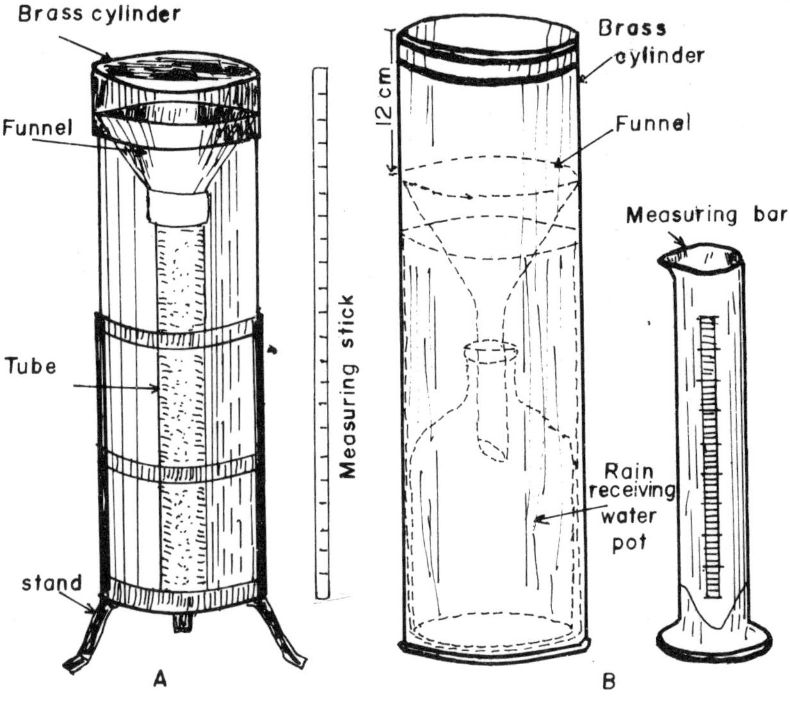

FIG. 4.7: Rain Gauge

minimize evaporation losses. Nowadays self recording gauges such as the hyetograph, tipping bucket and the float and syphon are used for accuracy. Similarly, where snow falls, snow gauges are used instead of raingauges.

Humidity

The amount of water vapour present in the atmosphere

It is a highly variable phenomenon, it continuously changes from place to place and from time to time. So it may be expressed in several different ways i.e. absolute humidity, relative humidity, specific humidity. Absolute humidity denotes the actual amounts of air, usually expressed in grams per cubic meter, whereas relative humidity is defined as the ratio between the amount of water vapour actually present in an air mass and the maximum amount that the air mass could hold at that temperature. It is always expressed in percentage. It is measured by means of hygrometer.

$$\text{Relative Humidity} = \frac{\text{Absolute Humidity}}{\text{Vapour Capacity}} \times 100$$

Saturated air has a relative humidity of 100%.

Specific humidity is defined as the ratio between the weight of water vapour (in grams) and the total weight (in kilograms) of a given volume of air. It varies from about 0.2 gm/kg in very dry cold arctic air to over 18.0 gm/kg in hot humid tropical air.

Hygrometer

It is an instrument that measures the relative humidity. It consists essentially of a human hair, which expands or contracts as the relative humidity increases or decreases, while another type uses a lithium chloride wire, the resistance of which varies with the changing humidity (Fig. 4.8).

A hygrograph is an automatic instrument which provides a continuous record showing changes of relative humidity over a given period of time. It is a type of hygrometer fitted with an inked pen resting on a rotating drum.

Psychrometer

It is also a type of hygrometer. It is used for measuring atmospheric humidity in which a current of air is artificially induced in order to produce the maximum evaporation at all times at the wet-bulb thermometer. It is almost totally horizontal movement of air over the surface of the earth. The speed and directions are measured by Anemometer and Wind Vane respectively.

Anemometer

It is an instrument that measures the velocity of the wind. It consists of four symmetrically mounted cups attached with arms (Fig. 4.9). These cups rotate freely on a vertical axis. The speed of rotation of the cups is recorded in miles, kilometers, or knots per hour by dial which is allocated at the bottom of the axis.

The other types of anemometer are the *pressure-tube anemometer,*

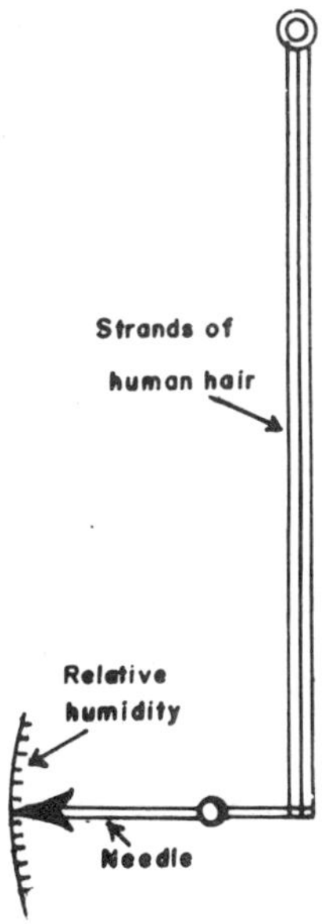

FIG. 4.8: Hygrometer

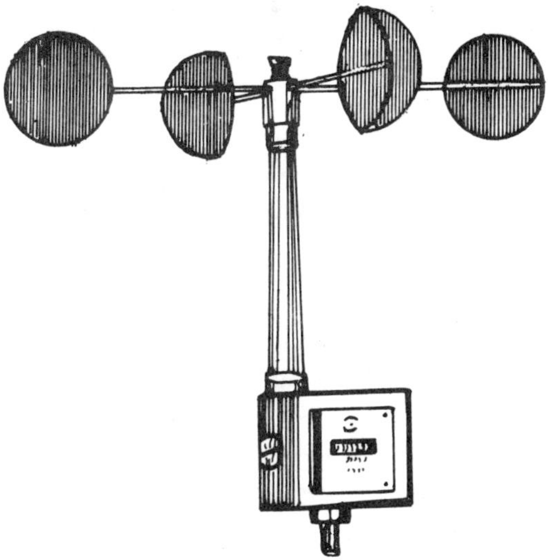

FIG. 4.9: Anemometer

in which the indication is the pressure built up in a tube that is closed at one end, the *pressure-place anemometer,* which measures the pressures exerted by the wind on metal plate, and the *hot-wire anemometer,* in which the current required to keep the temperature of a platinum wire constantly indicates the wind speed. An *anemograph* is an anemometer that automatically records wind speed and direction on a continuous trace.

Cloud

It is defined as a mass of minute water droplets, or ice crystals formed by the condensation of water vapour and held in suspension in the atmosphere. Condensation, which results from cooling, usually takes place around nucleus such as dust, smoke particles and salt ranging from ground level (fog and mist) upto over 13,000 m.

There are several classifications of clouds but by international agreement, clouds have been classified according to their height and appearance rather than their mode of origin. On the international system, there are four families and ten forms. The families and forms are listed below together with their general height.

Family A: High Clouds

(Mean Lower Level : 6,000 m or 20,000 ft)

1. Cirrus (Ci)

It is a high wispy cloud consisting of ice crystals at altitudes of 6000-12,000m. Due to higher altitude, the cloud is drawn out, forming mare's tails, which indicates strong winds in the upper atmosphere.

2. Cirrocumulus (Cc)

It is a combination of cirrus and cumulus. It is a type of high cloud (above 6000 m) consisting of ice crystals. It has a globular rippled appearance with blue sky showing between groups or lines of cloud. It is sometimes known as *mackerel* sky.

3. Cirrostratus(Cs)

It is a combination of cirrus and stratus. It is a thin high sheet cloud (above 6000 m) with a milky appearance. It generally casts a halo round the sun or moon and heralds a storm.

Family B: Middle Clouds

(Mean upper level: 6,000 m or 20,000 ft.)
(Mean lower level: 2,000 m or 9,500 ft.)

4. Altocumulus (Ac)

It is a fleecy cloud that occurs in bands or waves. Blue sky can usually be seen between the clouds although they sometimes join up. It occurs at about 2,400-6,000 m. It is usually a sign of fair weather.

5. Altostratus (As)

It is composed of either ice particles or water droplets or a mixture of both at altitudes of 2,000-6,000 m. It usually forms a greyish sheet, covering much of the sky. Precipitation is not common but it usually indicates rain to come as it heralds the approach of a warm front.

Family C: Low Clouds

(Mean upper level : 2,000 m. or 6500 ft)
(Mean lower level : Close to the ground)

6. Stratocumulus (Sc)

It usually forms a grey or whitish sheet or layer. It is recognised by large rounded masses or rolls visible within the general layer.

7. Stratus (St)

It is a low type of layer cloud with a fairly uniform base. It is generally grey coloured and is often thin enough for the rays of the sun to be seen through it. The cloud sometimes produces a light drizzle.

8. Nimbostratus (Ns)

It is a thick grey layer cloud that blots out the sun. It provides more or less continuous rain or snow falls. The base is low and uniform. It is most commonly associated with warm front of a depression.

Family D: Clouds with Vertical Development

(Mean upper level : 12,000 m. or 40,000 ft.)
(Mean lower level : 500 m or 1,600 ft.)

9. Cumulus (Cu)

It is often called cotton wool clouds. It is a cloud with a horizontal base and dome shaped upper surface (500-2000 m). It occurs mainly in summer due to convection. Sometimes it may cause thunderstorm, though ordinarily it is associated with fair weather.

10. Cumulonimbus (Cb)

It has a low base level (500-2000 m) but great vertical development, often reaching upto heights of 10000-12000 m. It is a development of the cumulus formation having anvil shape. Its presence indicates strong convection and there is frequently heavy precipitation and thunder.

Sunshine

It is that part of the sun's electromagnetic radiation which lies within the visible spectrum that reaches the earth's surface. It is measured by a sunshine recorder, which is called as the Campbell-Stokes Recorder. It can be measured either in hours per day or as a percentage of the total possible sunshine at the particular place. The duration can be as high as 80 per cent in cloudless tropical desert, or as low as 20 per cent in cloudy west coast temperate areas.

Campbell-Stokes Recorder

It is an instrument used for measuring and recording the duration of bright sunshine. As the sun moves round during the day, a spherical lens focuses its rays to burn a trace on a recording card. The length of the trace indicates the sunshine duration. The instrument has limitations in that it only records bright sunshine early morning and late evening sunshine may not be detected.

Pressure

Each gas has certain elastic force and weight. The elastic force at the surface is balanced by the weight of the column of air stretching from the surface to the higher limits of the atmosphere. Hence, the atmospheric pressure denotes either the elastic force or the weight of the air above a unit area of the earth surface. It is expressed in millibars (mb) and measured with the help of barometer. The mercurial barometer, the standard barometer, the Fortine Barometer, the Kew Pattern Barometer or the Aneroid Barometer. The average atmospheric pressure at sea level is 1013.25 mb. Atmospheric pressure decreases logarithmically with height.

Barometer

It is an instrument, used for measuring atmospheric pressure. The different types of barometer are described as follows.

Mercurial Barometer

It was invented in 1643 by Torricelli. It measures the height of a column of mercury in a vertical graduated glass tube, which is working on the principles of balancing (column of air and column of mercury). The tube is sealed from one side and filled with mercury and inverted into a glass cup which also contains mercury (Fig. 4.10). Adjustment need to be made for variations in gravity, temperature and altitude. Under standard condition (45°N and 12° C), the mercury is adjusted at 76 cms or 29.92 inches high in the tube at the sea level. Even slight fluctuations in air pressure will produce differences in the height of the mercury in the graduated tube. The pressure of a particular place is known as Station Pressure. Thus, the Station Pressure varies from place to place. Therefore, the Station Pressure is converted to sea level pressure in order to make a direct comparison.

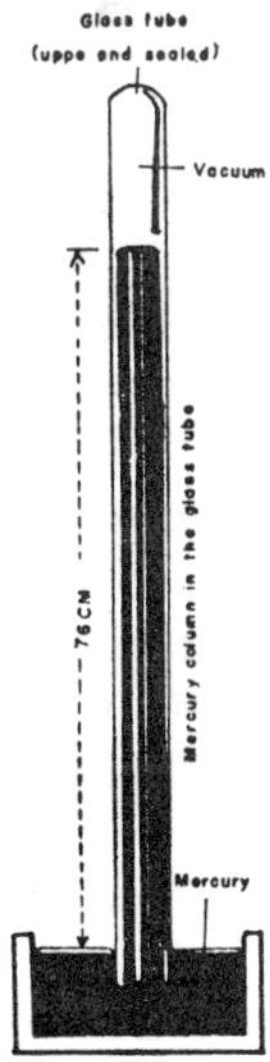

FIG. 4.10: Mercury Berometer

Aneroid Barometer

The name is derived from Greek word aneros, which means without liquid. It is light and portable instrument. It consists of a metal box from which the air is virtually exhausted. The sides are flexible, which expand and contract with changes in air pressure. The movements are amplified and registered on a caliberated circular dial with the help of a needle. It gives readings accurate upto ±1.5m. under favourable conditions (Fig. 4.11).

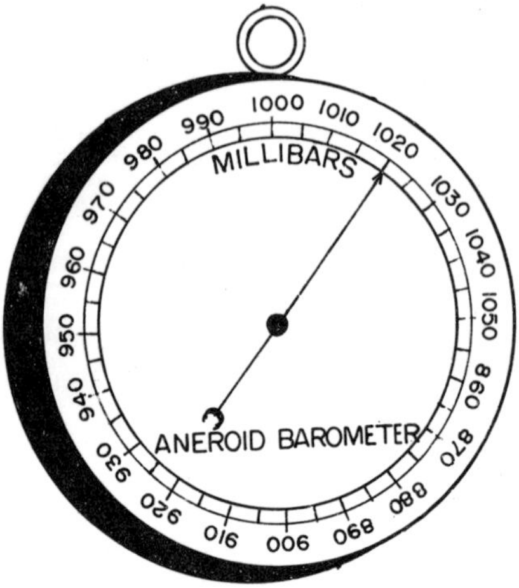

FIG. 4.11: Aneroid Barometer

Fortin's Barometer

It is little bit more sophisticated instrument used for measuring the atmospheric pressure. It consist of a glass tube, whose one end is closed and other is open. The tube is filled with murcury and attached to *cistern* from the open end side. The adjusting screw 'S' is fitted to *cistern,* which is used to bring the murcury level in the *cistern* to a fixed point. An ivory index 'I', which is fixed to the cistern, indicate the fixed point and zero of the scale (Fig. 4.12)

Barograph

It is an instrument which automatically records the continuous changes in pressure for a period of a week. Its function is similar to the thermograph. It comprises of a needle, caliberated chart and a rotating drum. Needle performs recording over rotating drum on which the chart is mounted (Fig. 4.13).

Pressure Altimeter

It is a form of aneroid barometer, which automatically measures changes in atmospheric pressure with height. It is used in aircrafts.

Isobaric System

Isobar is defined as line drawn on a map or chart joining places having the same atmospheric pressure. On weather charts, it is plotted to

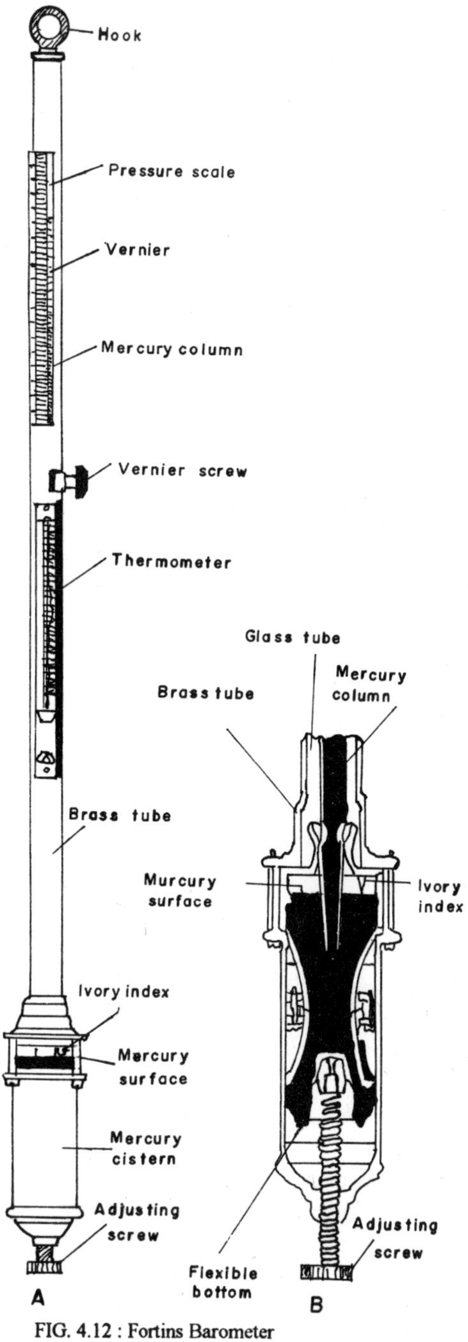

FIG. 4.12 : Fortins Barometer

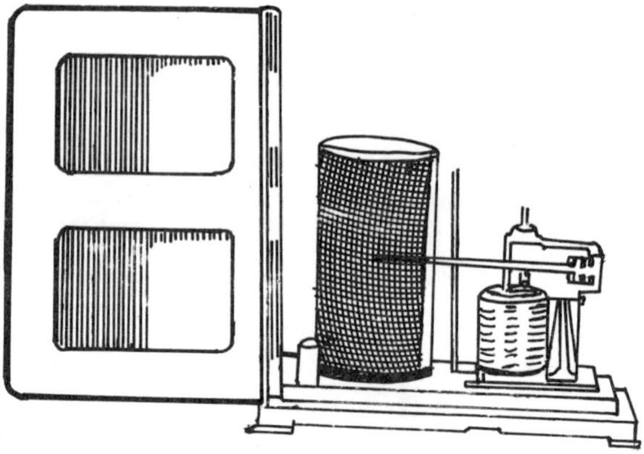

FIG. 4.13: Barograph

show the distribution of pressure at a particular time and is usually drawn at intervals of 2 or 4 millibars (mb). Some of the important systems are:

Cyclone

It is a system in which there is a low atmospheric pressure at the centre. It is a storm system of great intensity with winds circulating at speeds of 120-280 km per hour around the 'eye' (centre) of the cyclone, which is accompanied by torrential rain. Winds rotate in an anticlockwise direction, in the Northern hemisphere and in a clockwise direction in the Southern hemisphere.

Low or Depression

It is an area where the atmospheric pressure is lower than that of the surrounding area. It is found mainly in temperate regions. On a weather chart, it shows up as a series of enclosed isobars, more or less circular in shape, with pressure decreasing towards the centre. It can vary greatly in size (150-3000 km), speed (0 to 1000 km per day) and intensity. When the pressure difference between the centre and the surroundings is great, it is said to be deep, when the pressure difference is small it is said to be 'sallow'.

Anti-cyclone or High

It is an area of high atmospheric pressure. It is shown on a weather chart by a series of roughly circular closed isobars around the area of highest atmospheric pressure. Winds are light, circulate in a clockwise directions in the N. hemisphere and in anticlockwise directions in the S. hemisphere. Calm weather conditions are generally associated with anticyclones.

Col

It is an area of relatively uniform atmospheric pressure lying between two opposing anticyclones and two opposing depressions. The pressure being lower than that of the anticyclones and higher than that of the depression. The associated weather is variable with generally fine weather but frequent thunderstorms in summer and generally dull and foggy conditions in winter.

Ridge

It is defined as a narrow extension of an anticyclone, forming an area of high pressure, bounded on three sides by lower pressure areas. It often gives a brief period of fine anticyclonic weather in the midst of generally wet depressional conditions.

Secondary Depressions

It is a relatively small low pressure area that develops near a large depression. It frequently develops on the trailing cold front. It may become more intense than the large depression, resulting in heavy rainfall and stormy conditions.

Visibility

Visibility of air denotes the transparency of air. In other words, the greatest distance that an observer can see at a given time. It is usually quoted in kilometers or meters. It is largely dependent on the presence of particles (Water, ice, dust and smoke) in the atmosphere. It is reported on a scale ranging from 0 to 9 as given below.

Scale Number	Day light visibility
0	less than 50 m
1	50-200 m
2	200-500 m
3	500-1,000 m
4	1-2 km
5	2-4 km
6	4-10 km
7	10-20 km
8	20-50 km
9	Above 50 km

4.3 THE INDIAN WEATHER MAP

A weather map or weather chart shows on a flat sheet of paper on which all the data available concerning the weather for the given area at a particular moment of time is plotted. The elements of weather such as temperature, pressure, wind speed and direction, precipitation, fronts, clouds etc. are shown with the help of their respective symbols (Fig. 4.14) and these are constantly updated. The chart enables meteorologists and weather forecasters to observe the development of the weather patterns and to anticipate future developments. Conditions of weather elements are observed at different meteorological stations and sent telegraphically to the head office for the final preparation of the weather report.

The first publication of weather map was started at Simla in 1864. After First World War, the office of the Indian Meteorological Department was shifted to Pune. Till 1949, there were limited informations but after that a sufficient amount of informations used to be published. At present there are 350 meteorological observatories in India, which have been classified into various categories based on the size, facilities and nature of information.

4.4 ELEMENTS OF WEATHER

Information given

The Indian weather Map provides three charts in one sheet. There is one large map at the top accompanied by two small maps at the

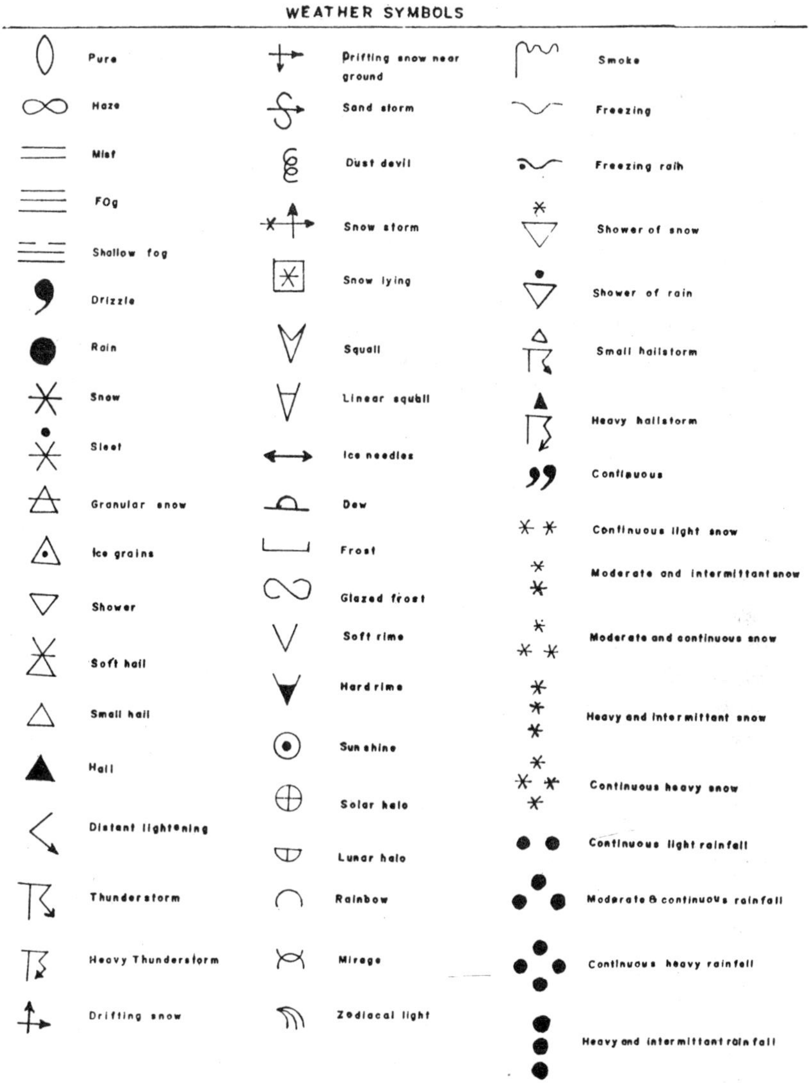

FIG. 4.14: Conventional sign.

bottom, which provide information regarding departure of minimum temperature from normal and the departure of 8.30 hours pressure from normal. On the back side of the chart there is a summary of the weather conditions and weather forecasting.

Weather elements, which are shown by large map are as follows:

1. Pressure

It is shown by means of isobars. The values of isobars are indicated in the force unit of millibars (mb). The areas of Low pressure and High pressure are marked on the chart as 'L' and 'H' respectively.

2. Wind

Wind direction is shown by means of an arrow. The wind speed is shown with feathers at the back of the arrow. The length as well as thickness of the feather varies according to the wind speed. The wind speed and directions are shown for the station where cloud cover exists.

3. Cloud

Cloudiness at a station is shown by means of shading small circles at the station. The circle has been divided into 8 equal parts. The amount of clouds are determined by counting the shaded parts of the circle. On the chart only medium and low clouds are shown. In the international map, circle has been divided into 10 equal parts. An overcast sky is shown by completely shaded circle.

4. Rainfall

It is shown through the numerical values and plotted on the map outside the circle. The amount of rainfall below 44 cm is shown by the symbol.

5. Sea Condition

It is depicted through waving arrow along with the alphabetical symbol. The alphabetical symbols are as follows:

Symbol	Sea Condition
Cm	Calm
Sm	Smooth
Sl	Slight
Mod.	Moderate
Ro.	Rough
V.Ro.	Very Rough
Hi.	High
V.Hi.	Very High
Ph.	Phenomenal

4.5 STUDY OF THE INDIAN DAILY WEATHER REPORT

Various symbols are used to represent haze, mist, lightning, dust storm, shallow fog, snow, squall, drizzle, shower, thunderstorm, hail etc. around the station circle.

The weather information can be studied systematically under the following heads.

1. Introduction

Day, date, time and season

2. Pressure

(i) Low pressure (L)
(ii) High pressure (H)
(iii) Isobaric trend
(iv) Press gradient (steep and gentle)
(v) Depression and secondary depression

3. Wind

Direction and Velocity.

4. Rainfall

Heavy, low or general

5. Cloud

(i) Sky cover
(ii) Type of clouds (Low, middle or high)

6. Other Phenomena

7. Temperature Departure from Normal

8. Pressure Departure from Normal

9. Sea condition

Example: Interpret the weather information available on the weather map at 08.30 Hrs. IST (03.00 G.M.T.) Tuesday, the September 11, 1951.

Introduction

The given weather Map provides the information about the Weather conditions on Tuesday, the September 1951 at 08.30 Hrs. Indian Standard Time (0300 Greenwich Mean time). In September, India is under the influence of south-west monsoon and it is a time of general rains. This weather map provides the report dealing with the weather conditions of Pakistan, Sri Lanka (then Ceylon) and Burma (Myanmar) also on the above given date (i.e. 11th Sept.). The given figure excludes (Myanmar) and shows Pakistan, Sri Lanka and India only.

Distribution of Pressure

Over the Indian Sub-continent, two areas of low pressure exist, one of which extends over western parts of Sind, southern part of Punjab (Pakistan) and most of Baluchistan. Monsoonal winds are attracted by this low pressure area and the pressure in this particular area is 1004 mb in the given weather map. The other low pressure area coasts over east Uttar Pradesh and the adjoining parts of Bihar and Madhya Pradesh. To the south west of Mangalore over the Indian ocean is high pressure area. This high pressure area penetrates southern Ceylon but it is not deep and is represented by a wedge of 1012 mb.

Trend of Isobars

This High Pressure affects the trend of isobars on the west coast of the Indian Peninsula. The isobar of 1012 mb sweeps north and then descends in a north-south direction. The isobars of 1010 mb, 1008 mb and 1006 mb also follow the same trend and curve towards the north, although their curvature is slightly smaller as compared to the 1012mb isobar. The isobars which till now curved to the north, now describe the earlier trend of curving towards the north. The presence of low pressure area over east Uttar Pradesh and the adjoining parts of Bihar and Madhya Pradesh determine this trend of curvature of isobars.

Pressure Gradient

The closeness or otherwise of the isobars determines the pressure gradient. Pressure gradient is very gentle over Punjab, most of the western U.P. and over north-eastern Rajasthan. Isobars are absent between the low pressure area in the west and depression over east Utter Pradesh and its adjoining areas. The isobar of 1004 millibars bound both these lows. Over the peninsula, the pressure gradient is again gentle. As the isobars have come closer, the gradient is somewhat steeper over the western and northern fringes of the latter depression.

Wind

Over rest of the country, the pressure gradient is gentle. Therefore, the wind velocity is slow. It is generally 5 knots per hour, but in some places the wind velocity is more. To quote few examples at isolated stations in Madhya Pradesh and Orissa and over the sea, the wind velocity is more. The wind direction is from the west to the east over the peninsular northern Bombay, Saurashtra and western Madhya Pradesh. The direction of the wind over Bengal, Assam, Bihar, Madhya Pradesh and U.P. is under the influence of the cyclone that exists there. The wind is moving in anticlockwise direction round the cyclone and as a result it blows from the south-east over Bengal and Bihar, from the east over northern Bihar and northern U.P. and from the south over eastern Madhya Pradesh.

Cloudiness

The cloud cover is widespread over the northern west coast, southern peninsula, north eastern coast, Bengal (east Pakistan, now Bangladesh), West Bengal, Assam, Bihar, U.P., and eastern Punjab. Low clouds make the sky overcast over east Bengal (now Bangladesh), West Bengal, Assam, Bihar and U.P. Over Bihar and east U.P. also, high clouds occur. Half or more than half of the sky is covered by clouds over the states of Maharashtra and Gujarat. High or Medium clouds occur over most of the sky over the peninsula. Less than 1/4 of the sky is covered by clouds in Rajasthan, Punjab and most of western Pakistan.

Other Weather Phenomena

Drizzle is going on at Tezpur and Mangalore and a thunderstorm is visiting Dibrugarh.

Temperature Departure from Normal

Temperature departures are very slight over most of the country. Two important exceptions are over south-east Madras and east Assam, where the minimum temperature is 4 degree Fahrenheit more than the normal and over Bihar, where temperature is 4 degree less than the normal.

Pressure Departure from Normal

Due to the presence of the depression over former Vindhyachal Pradesh, Bihar and U.P., the pressure is less by 2 to 4 millibars. The pressure is slightly above normal i.e. two millibars more than the normal over the rest of the country.

Sea Condition

Weather chart does not provide us information about the sea condition on the particular date.

5

GEOLOGICAL MAPS

The different rock-types of the area are displayed by the help of a geological map, which is superimposed upon its topography. These types of maps explain the information concerning the geological structure of the region. In brief, a geological map offers all possible data about the geology of the area concerned.

5.1 GEOLOGICAL MAPS

There are mainly four types of geological maps.

(i) Surficial maps
(ii) Outcrop maps
(iii) Areal maps
(iv) Structural maps.

Surficial maps are those which show the features of the surface of the earth e.g. Agricultural soil. Geological categories of this type of maps i.e. stream gravel, stream alluvium, glacial gravel etc. are very useful from engineering point of view especially in the construction of irrigation channel, airports etc.

Outcrop maps show the area of bed rocks, which due to erosion or weathering are exposed to surface of the earth. Bedrocks are those rocks, which are situated just below the earth surface.

Areal maps explain the plain view of the geological structure of a particular area. It is mostly used in a physical geography and Geomorphology.

Structural maps are like the areal maps. The basic difference is that, in structural maps conventional structural symbols are used. It is much more informative than areal maps.

Some scholars have divided geological maps into two groups.

1. Geological drift maps.
2. Solid geological maps.

The main feature of a geological map, which attracts immediate attention, is the display of symbols or colours indicating the location and extent of the different rocktypes of the area. The outcrop of a rock body is the intersection of the same with the topography of the region. Thus, for a bed, 100 feet thick and dipping at an angle of 30 degree towards east, the outcrop on a level surface would run north-south and would have a width of 200 ft. (Fig. 5.1).

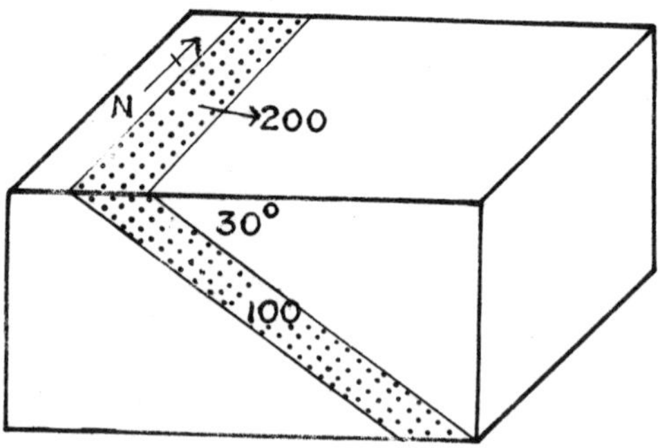

FIG. 5.1: Geological drift map.

The dip of a bed is its amount of inclination with respect to the horizon, measured along the line of greatest slope of the bed. This is given in degrees. This is always at right angles to the strike. For a horizontal bed, the dip is 0° and for a vertical bed, 90°. The geographical direction in which a bed has maximum slope is known as the direction of dip. The direction of the line along which the inclined bed intersects a horizontal plane is known as the strike of the bed. It is the direction, along which the bed appears to be horizontal. In between the outcrops of any two of the beds in a map, there lies always a boundary line separating the two. This line is designated as the strike line. In other words it is also known as bedding plane.

If all the boundary surfaces i.e. all the beds dip in the same direction and at the same angle, the beds are mutually conformable. If the case is not same, then there is some criteria to determine the beds.

5.2 CRITERIA FOR UNCONFORMITIES

When the two series of beds belong to two different geological ages and they are, in most cases, different in their geological structure, the following relation generally holds good.

(a) Some of the beds appear to cover up some other beds (the bed A, in Fig. 5.2)
(b) The boundaries of the latter appear to end abruptly against that of the former (Fig. 5.2).
(c) The dips of the beds differ in the two different formations thus related.

The plane that separates the two series of beds of two different geological ages is known as plane of unconformity i.e. (X-X' in Fig. 5.2).

5.3 CRITERIA FOR FOLDING

In case of folding, the fold may either be a syncline or an anticline or assemblage together (Fig. 5.3). In this case:

(a) The outcrop of a bed generally repeats itself several times.
(b) It along with (a), the bed is found to dip in opposite directions in any two successive outcrops, it is definitely a case of folding (Fig. 5.3).

5.4 CRITERIA FOR FAULTS

In case, a fault is present in a map, generally the line of fault is marked on the map, and the fault is vertical or inclined accordingly as the fault line is straight or curved (Fig. 5.4, X-X is the fault lines).

5.5 THICKNESS OF BEDS

In general the accurate thickness of a bed can be determined by drawing a section through the outcrop of that bed, along the direction of true

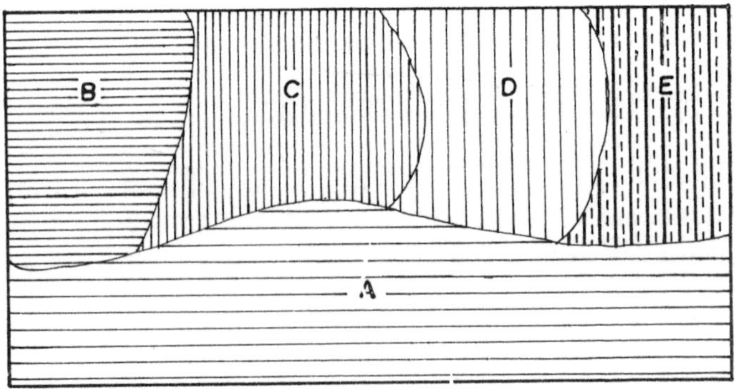

FIG. 5.2: Map to show unconfirmity

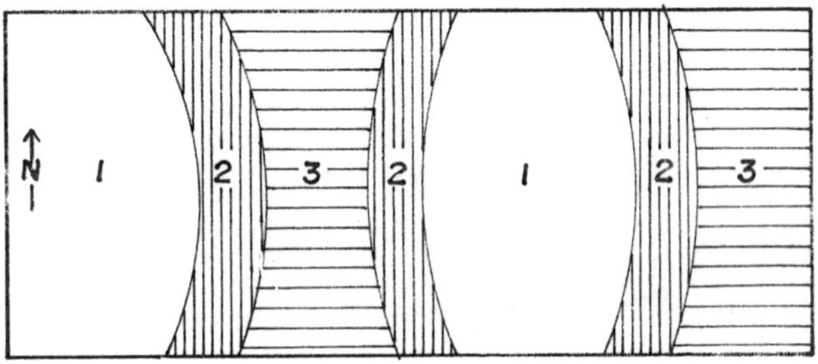

FIG. 5.3: Folding map

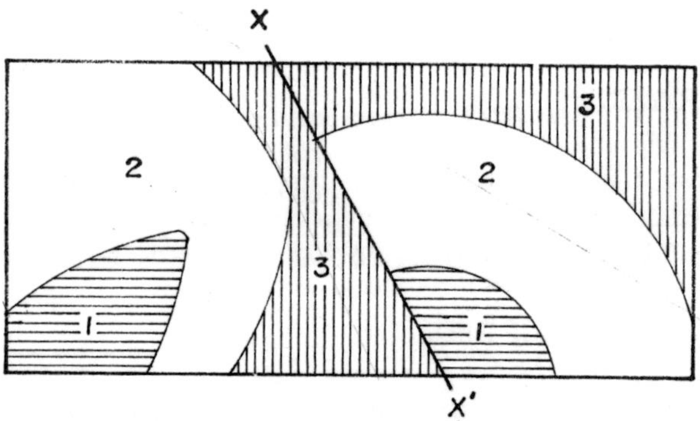

FIG. 5.4: Faulting map

dip and measure of the thickness of bed from the section, drawn according to scale. This is true thickness of the bed.

How to draw a section of a map along a given line.

(a) Mark the line of section on the map in such a manner that it passes through all the beds and draw a horizontal line of the same length which becomes the base line of the section to be drawn (Fig. 5.5).

(b) Mark the points of intersection of the line of section on the map *(A-B)* with different contour lines and transfer these points to the base line of the section (points *C, D, E, F, & G)*.

(c) At those points on the base line, draw perpendicular line and cut off lengths proportional to the corresponding contours, according to scale.

(d) These points on the perpendicular line should then be marked distinctly and the perpendicular line should be erased. The points should now be joined with a free-hand to construct the surface profile of the region, along the line of section (*XY* is the profile).

(e) Now, the points of intersection of the boundary surfaces between the different beds with the line of section should be transferred to the base line and then projected to surface profile (*M* and *N* are the points).

(f) From these points, line should be drawn in accordance with the dip of the particular boundary surface under consideration. To find out the dip, draw strike line and take the distance between them. Mark their values at both ends and draw a perpendicular line at the higher value points according to the scale given on the map i.e. if the contour interval is 100' and scale is 1" = 1000' then perpendicular line is.

Here 1000', is represented by 1"

$\therefore$ 1, is represented by 1/1000"

$\therefore$ 100 is represented by 1 × 100/1000"

= 0.1

Thus dip angle is: 134°.

(g) When the section is complete, it should be coloured or shaded in accordance with the colours or shades respectively in the map. The vertical and horizontal scale, dip angle and index should always be mentioned below the section. After

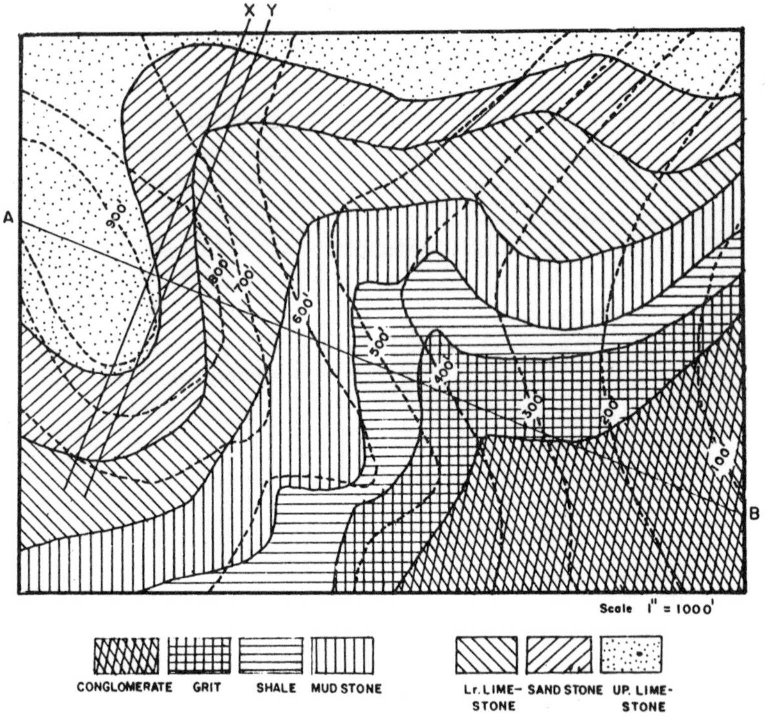

GEOLOGICAL SECTION

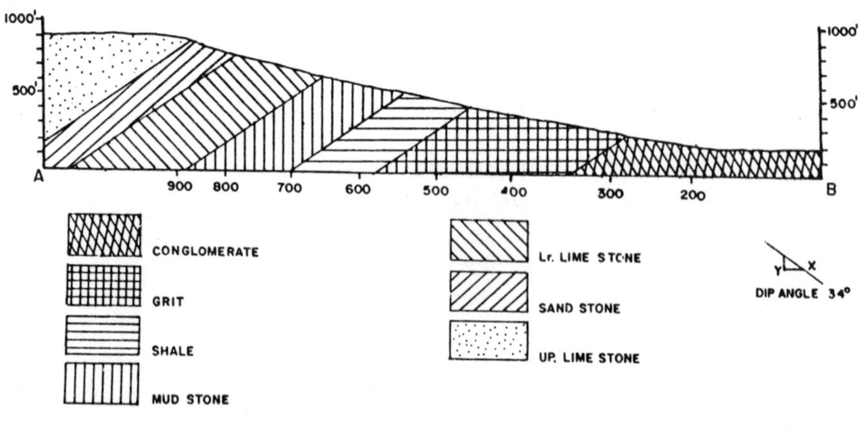

FIG. 5.5: Geological map and its section

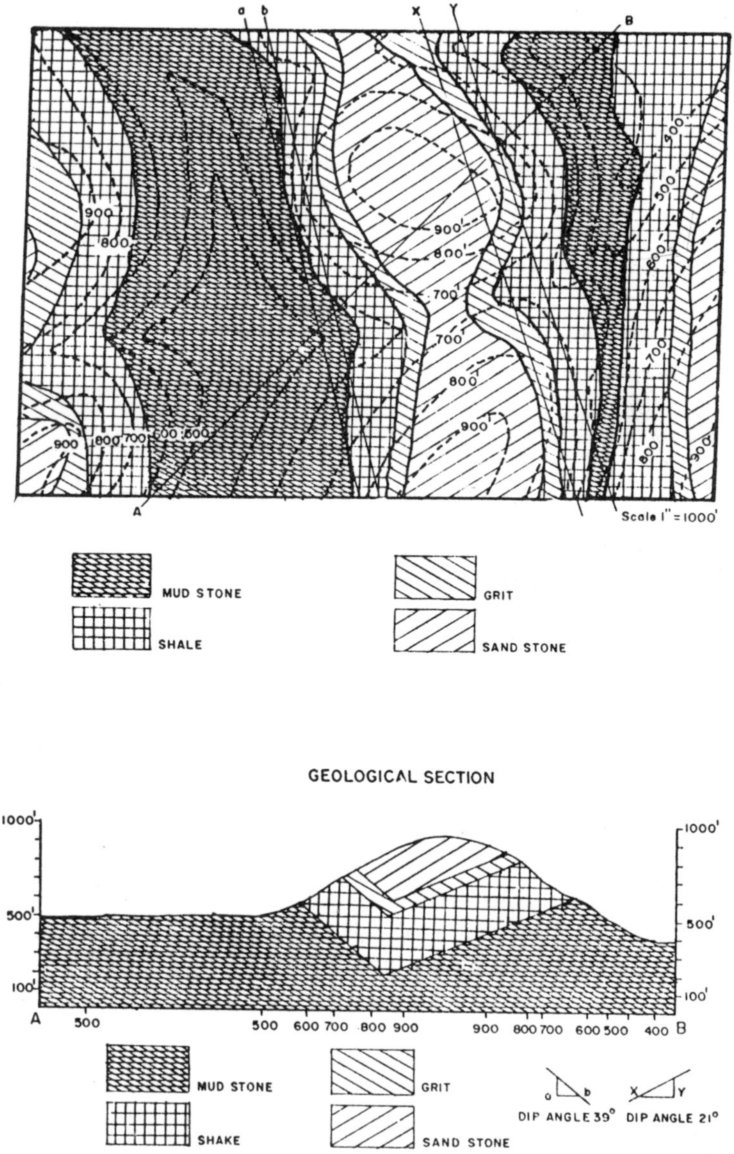

FIG. 5.6: Geological map and its section

completing the section, write in brief the geological history of the region. Under this heading give the order in which the beds were formed and give in chronological order, the sequence of events like folding, faulting etc.

5.6 EXERCISES OF GEOLOGICAL MAPS AND ITS INTERPRETATION

Some exercises of geological maps of different nature and their interpretations are given as follows:

Example: Draw a geological section and interpret the given geological map (Fig. 5.5 and Fig. 5.6).

Geological History

The beds are formed in order, under sedimanetary conditions. They are then subjected to compaction and upleft, tilting. The oldest or the lower most rock of the region is conglomerate and the upper most or the youngest rock is upper limestone. In between these from older to younger are Grit, Shale, Mudstone, limestone and sand stone. The contour interval of this region is about 100 feet. The thickness of the bed varies from 75 feet to more than 500 feet.

Geological History

The map shows that there are the repetition of beds, which is the sign of folding. It means, the beds are folded due to comparational force. The countour interval is of 100 feet. The appearance of contours show that the region is very much undulating having one pronounced anticline surounded by 900 feet contour. The oldest rock is mudstone and the youngest is sandstone. In between are shale and grit.

6

SLOPES AND GRADIENTS

Slope is the steepness of a land surface. Slopes play an important role in the formation of a landscape. Various methods are in use to express slope conditions. The most obvious method is to express slope on some kind of map by contours. Because, it is easy to find out the nature of slope by contour map. Contours further apart represent a gentle slope and on the other hand contours closer to each other represent steep slope. Overall, slopes are of greater use.

6.1 KINDS OF SLOPES

(1) Uniform
(2) Convex
(3) Concave
(4) Undulating

1. Uniform

In a uniform slope, the distance between contour lines remains the same throughout the region (Fig. 6.1).

2. Convex

A convex slope is marked by an outward bulge. The slope rises more steeply at the lower ground as compared to the top. On a map it is shown by closely spaced contours at the foot and comparatively further apart at the top (Fig. 6.2).

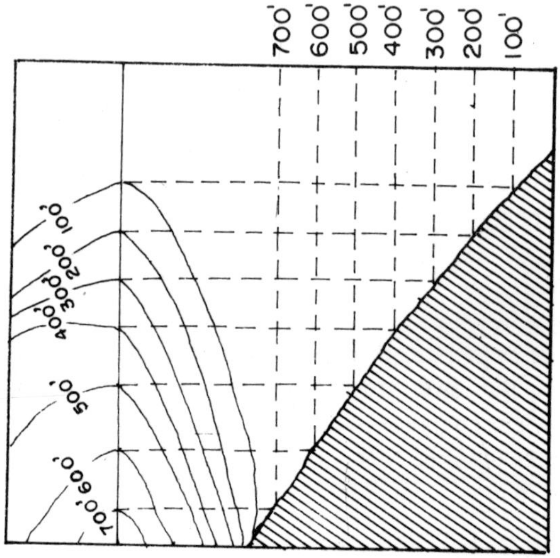

FIG. 6.1: Uniform slope

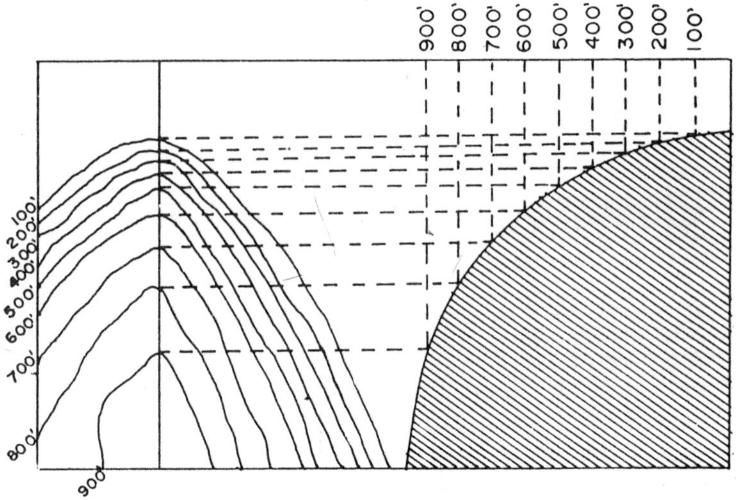

FIG. 6.2: Convex slope

3. Concave

Its shape is just opposite to convex slope. The slope is less at the bottom but more at the top. On a map it is shown by closely spaced contours at the top and comparatively further apart at the bottom (Fig. 6.3).

4. Undulating

It shows the combination of concave and convex slopes. On the map, it is shown by the contours, which have no uniform spacing (Fig. 6.4).

Several methods are used to express the slope. The important one are as follows:

(a) Degree
(b) Gradient
(c) Percent
(d) Mills

Degree

Expression of slope by degree can be obtained by two methods.

(i) By the help of protractor
(ii) Trigonometrically

Suppose a right angle triangle ABC (Fig. 6.5) in which BC represents the slope distance. ABC represents the slope degree, which can be measured by the help of protractor.

Trigonometrically *BC* can be calculated by the following formula;

$$\tan\theta\,(ABC) = V.I/H.E$$
$$= 5/10 \text{ (Supposed values)}$$
$$= 0.5$$
$$\text{Tan}\,\theta = 0.5$$
$$\therefore \theta = \text{Tan}^{-1}\,0.5$$
$$= 26.5^{o}$$

Gradient

It is defined as the ratio between vertical height (V.I.) and horizontal

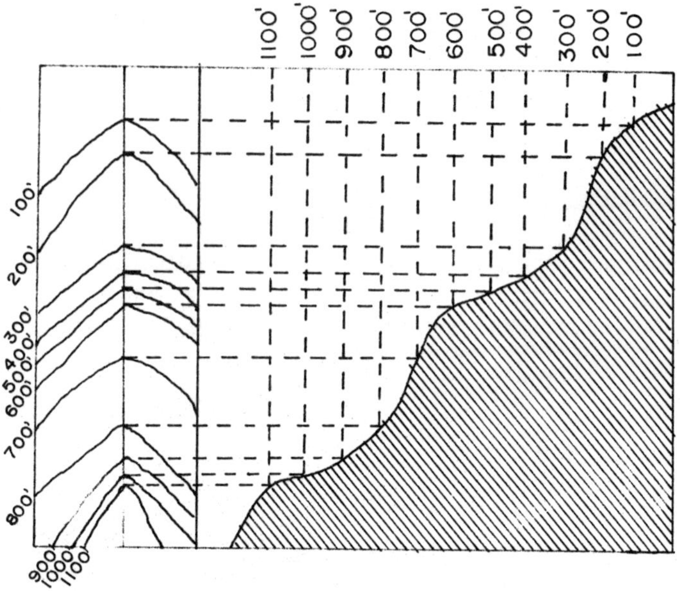

FIG. 6.4: Undulating slope

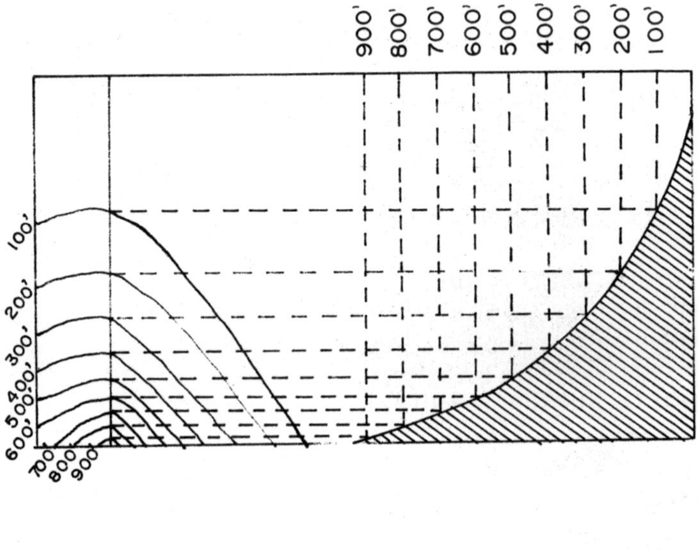

FIG. 6.3: Concave slope

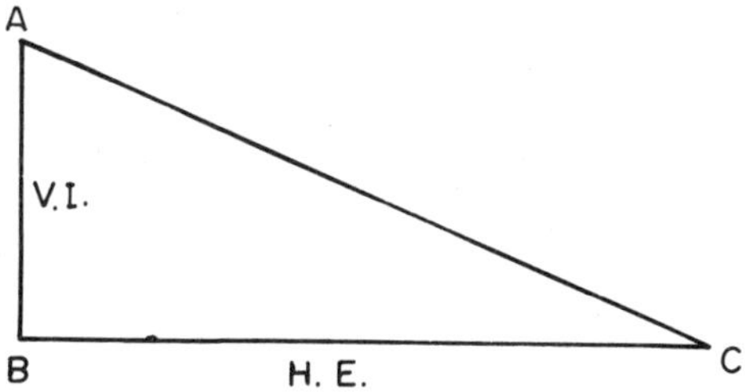

FIG. 6.5: Expression of slope by degree

distance (H.E) and expressed in terms of fraction in which numerator is unity. If V.I. = 5 ft. and H.E. = 10 ft., the gradient will be 5/10 = 1/2 which indicates a rise of 1 ft. after 2 ft. horizontal distance.

Per cent

The slope is determined by the help of triangulation method and it is transformed into percentage by multiplying with 100.

For example;

Percentage of slope = V.I/H.E × 100

Mills

It is specially used by army. Its unit of measurement is a 'mil'. The easy way to get the mills is to multiply the percent slope by 10. In other words multiply the vertical interval (V.I) by 1000 and divide by the horizontal distance (H.E).

Slope in Mills = V.I. × 1000/H.E

or Slope in Mills = Percent slope × 10

6.2 METHODS OF REPRESENTING RELIEF

There are several methods which have been applied in the slope analysis of maps, the important ones are as follows:

In 1890 S. Finster Walder and K. Peucker devised a method first time for determining the average slope of land surface. This involved measuring the total length of all contours with an opisometer, and applying the formula.

$$\text{Sine of degrees of average slope} = \frac{\text{Total length of contours} \times \text{Contour interval}}{\text{total area}}$$

(i) C.K. Wentworth's Method of Average Slope

In 1930 C.K. Wentworth developed a simplified method for determining the average slope of land surface. He devised a 'general and random' method of determining average slope over an area from a map. According to this method firstly the contour map of the area is covered with an east-west, north-south grid, then contour crossings are counted and then tabulated for determining the average number of contour crossing per mile. The procedure was repeated using an oblique grid over the same area and the average number of contour crossings per mile determined. He then applied the following formula to determine the tangent of the average angle of slope of the land surface.

$$\text{The average angle of slope } (\tan\theta) = \frac{\text{Average no. of contour crossings per mile(A)} \times \text{contour interval (I)}}{3361 \text{ (Constant)}}$$

The result gave the average slope in terms of the tangent, which can be converted into degrees with tangent tables.

(ii) G.H. Smith's Method of Slope Analysis

According to this method, the topological sheet is divided into rectangles of 5 minutes each by longitude and latitude. Then the difference in height between the highest and lowest point in each rectangle is calculated and plotted in the centre of each square on a small scale base map. Isopleths are drawn for each 100 ft. to indicate areas having the same amplitude. The map is then shaded in different tints to indicate areas with same 'local relief'. Further information is obtained by measuring the area of each 'relief province' and then expressing each area as a percentage of the total land area. This method is effective in maturely dissected plains and plateaus, but does not provide satisfactory results in mapping the complex topography.

(iii) The Raisz and Henry Method of Average Slope

Raisz and Henry tried to improve the Smith's method by dividing the topographical map into unequal rectangles with some physiographic identity i.e. incised valleys, monadnocks etc. But the unequal areas proved unsatisfactory. Finally, they divided the large scale topographical map into small regions, within each of which the contour lines had the same number per mile of horizontal equivalent. Seven categories were selected representing slopes of under 50 feet per mile, 50-100 feet, 200-300 feet, 300-400 feet, 400-500 feet and above 500 feet. A horizontal scale of standard contour spacings was drawn to check the number of contours per mile by the help of dividers and carefully ascertained the slope categories. After the categorical demarcation on the large scale maps, their boundaries were transferred to the small scale state map, and tinted in seven shades.

(iv) A.H. Robinson's Method of Slope Analysis

A.H. Robinson devised a quantitative accurate relief map. According to this method, a map should be covered with a network of squares each having 0.01 square miles in area. The average slope of each square was determined and marked on it by placing one dot for each 1 degree or ½ degree of average slope. The dots were not placed symmetrically but its position were determined according to the contours on the map and to the dots in the adjacent squares. Thus a dot map of average slope is prepared.

6.3 RELIEF REPRESENTATION

Relief indicates the general features of the land surface i.e. relative height and low lands. It can be represented by three methods.

(i) Pictorial Methods

(a) Hachures

This method was invented by Major Lehman. Hachures are the broken line, sketched on maps to give an impression of the relief. The lines, run down the slopes and are closer together where the slopes are steeper and they are thinner where slopes are gentler. So the steeper the slope, the darker the shade. Above 45 degree they are shown completely black. Flat areas with slight slope, tops of the mountains and valleys are left completely white. It is thus clear that the Hachures fail to show the exact slope (Fig. 6.6).

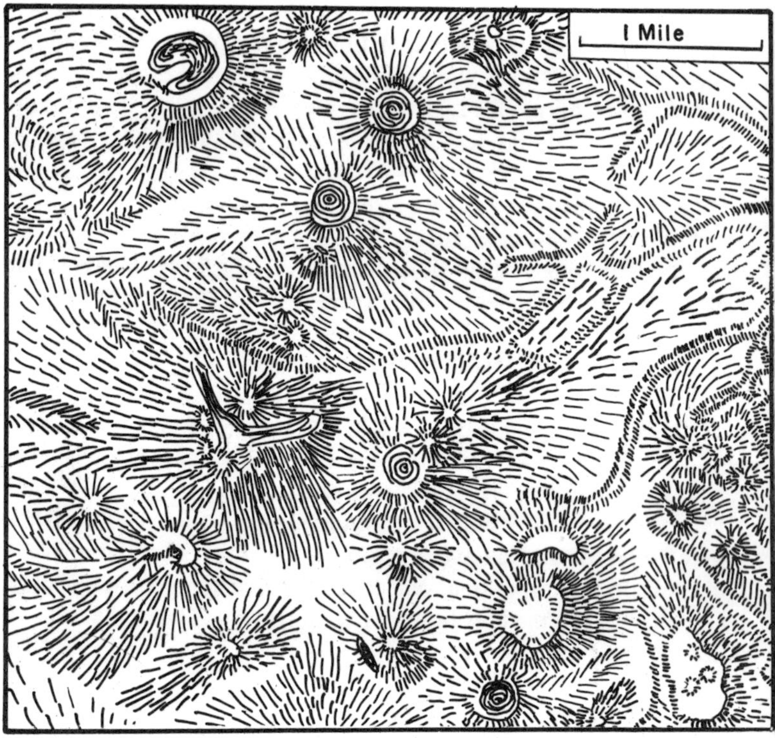

FIG. 6.6: Hachures and Hill shading

(b) Hill Shading

A rounded elevation of the earth's surface, which is of lower altitude than a mountain having less steeply inclined sides. It is known as "plastic shading" in U.S.A. It aims at producing something of the effect of a relief model. The shading is done by dots. The method is similar to hachuring. In this method light is thrown either vertically or obliquely, depending upon the location of the slope side of the region. It provides, however, some idea about the relief of the region or country.

It is generally used in small scale maps (Fig. 6.6).

(ii) Mathematical Methods

(a) Spot Height

The height, that has been precisely measured above mean sea level on the surface and shown on map by dots with the heights in figures. Thus, these do not indicate the relief of the land by themselves (Fig. 6.7).

(b) Bench Marks

A mark made by surveyors to record a point of known position and height above mean sea level. The mark is cut into a permanent feature. It consists of a broad arrow with a horizontal bar through its apex. The arrow points upwards unless the location is below sea level. On the maps these are indicated by the letters B.M. and the height in figure. (Fig. 6.7)

(c) Trignometrical Stations

These are indicated on maps by a small triangle with the heights in figures above sea level. These points are used for triangulation survey.

(d) Form Lines

These are the approximate contours, which represent the minor details of the topography. These are shown by broken lines and drawn in between the instrumentally fixed contours having large contour intervals.

(e) Contours

Contouring is the most important and standard method of representing relief. Contours are imaginary lines drawn on a map joining all places with the same height above and below mean sea level, which are fixed by accurate survey. The contours are drawn on certain intervals i.e. 10 ft to 20 ft etc. which depends upon the scale of the map. A contour map also informs about patterns and relative slope of the region. Wherever the contours are closely spaced, the slope is steep, wherever they are far apart, the slope is gentle (Fig. 6.7).

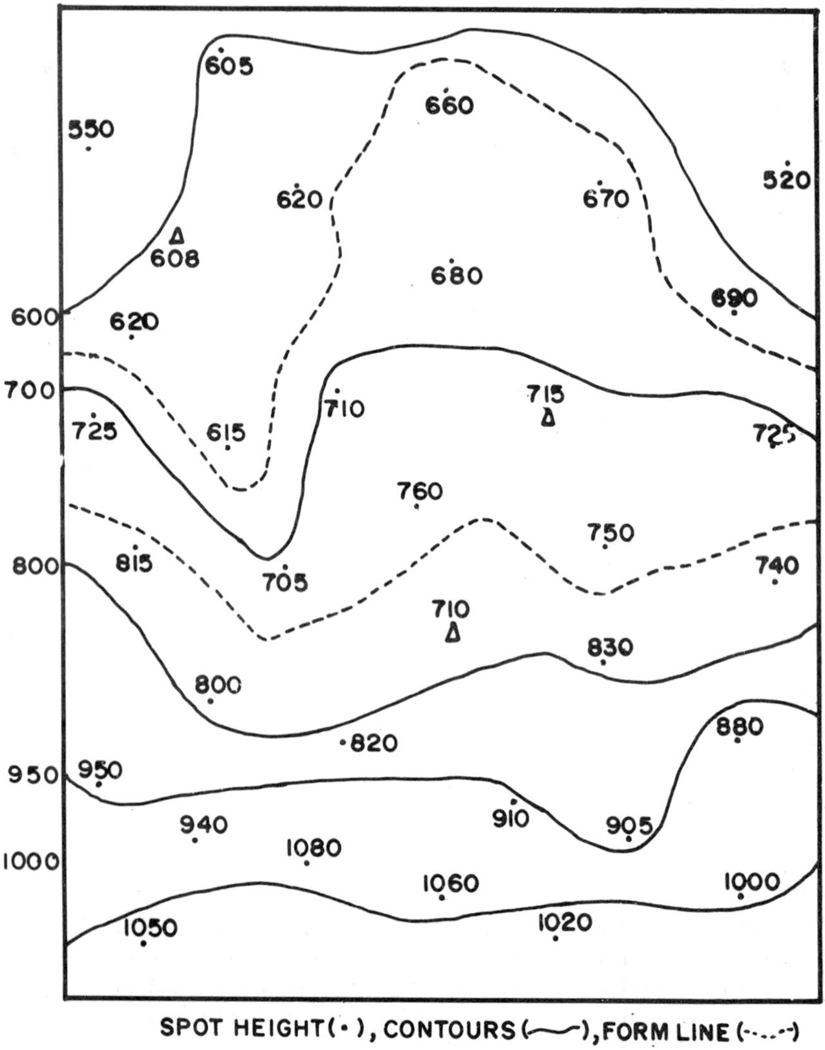

FIG. 6.7: Contours, Formlines, Trignometrical stations and spot height

(iii) Combination of Both

The above mentioned methods are suffering from certain drawbacks, therefore, for better and sufficient representation, a combination of different methods provide best results. The combination may be as follows:

(a) Contours with Hachures
(b) Contours with Plastic Shading
(c) Contours with Hachures with Plastic Shading
(d) Contours with Spot Heights
(e) Contours with Form lines with Spot Heights
(f) Contours with Hills Shading etc.

6.4 RELIEF REPRESENTATION BY CONTOURS

Contouring is one of the most important method to represent the relief features. It provides two-dimensional view of relief. Some of the important tips for drawing a cross-section are as follows:

Example: Draw a cross-section by the help of the following contours.

Firstly, select a scale. Draw a line AB in such a way that it cuts almost all the contours and then draw a parallel line A'B' to AB. Construct a scale (on your choice) on line AA' and BB' from points A' and B' onwards. Take those points which are cut by the section line (AB) on contours. From those points, draw a perpendicular line upto its horizontal scale. Join the meeting point of horizontal as well as perpendicular line by free hand smoothly (Fig. 6.8).

Some of the common relief features and their cross-sections are given below:

Hill

A rounded elevation of the earth's surface, which is of lower altitude than mountain (generally less than 3,000 ft) and with less steeply inclined sides. The spacing of the contour indicates the character of the hill, that is, whether it is conical or flat topped etc. In case of conical hill with uniform slope, hills are represented by almost circular and regular contours (Fig. 6.9).

Mountain

An upward projection of the earth's surface that rises to high altitudes.

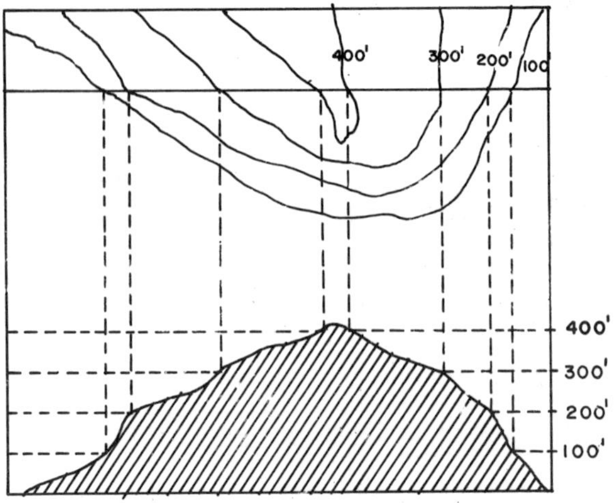

FIG. 6.8: Relief representation by contours

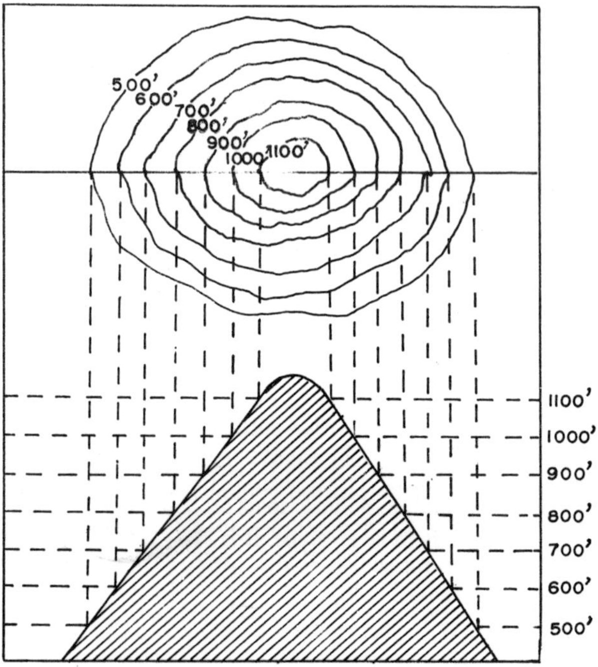

FIG. 6.9: Representation of Hill

It usually possesses steep slopes with sharp ridges and one or more rocky peaks. The contours help to show the nature of the mountains, that is, whether it is narrow or flat topped etc. (Fig. 6.10).

Ridge

It is an elongated hill or narrow range of hills having steep slope. A ridge generally connects two hills. It is represented by elliptical contours. The line, which represents greatest elevation on the ridge is called the ridge line (Fig. 6.10).

Saddle

It is a broad shallow gap in a mountain ridge. It is represented by two sets of circular contours apart from each other and further, surrounded by elliptical shape contours (Fig. 6.10).

Plateau

It is an extensive elevated area of relatively flat land having steep slopes atleast on one side. It is usually called a table-land. It is represented by close contours of square or rectangular shape having broad centre (Fig. 6.11).

Knoll

It is a small rounded hill or mound, which is often crowned with trees. If it is prominent, it is called hillock. A slight elevation of an irregular shape is called a rise. Mostly it is represented by a concentric contour (Fig. 6.12).

Spur

It is a projecting part of a hill, especially a convex portion of hill side between two hollows. Shorts spurs are known as bultressers or shoulders. If the spur ends in steep slope, it is known as *bluff*. It is represented by the contour bulging outward in which higher contour bends toward a lower contour (Fig. 6.13).

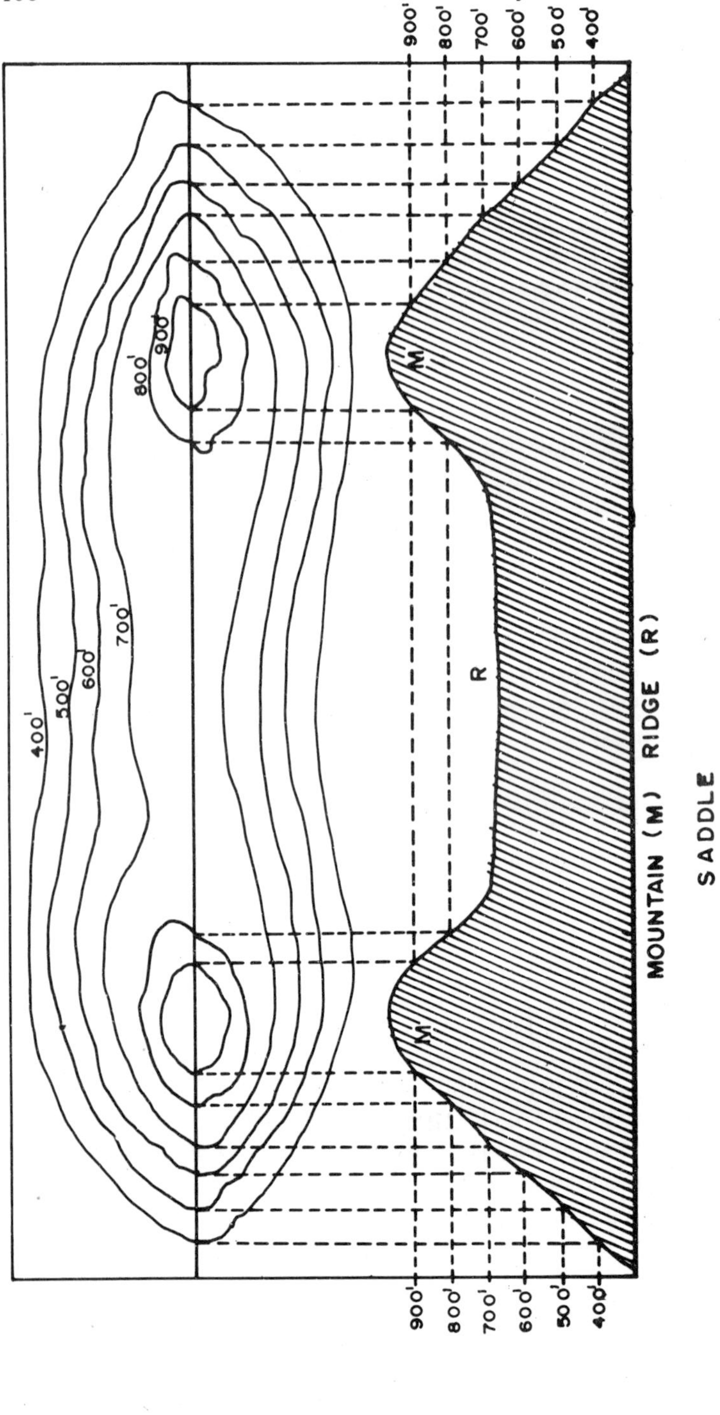

FIG. 6.10: Representation of Mountain ridge and saddler

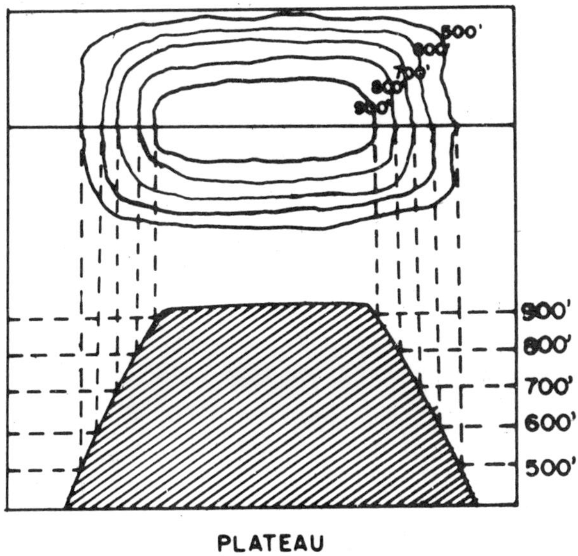

FIG. 6.11: Representation of plateau

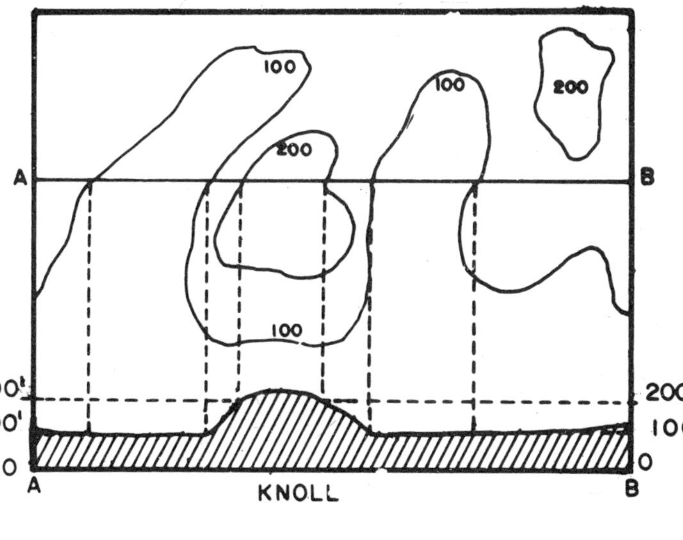

FIG. 6.12: Representation of knoll

Col/Pass

It is a routeway between two mountain summits at a high level, formed by glacial or fluvial action. It is often of significance for communications, providing natural routes through mountain regions. It is represented more or less like saddle. But the circular contours are comparatively closer in case of col/pass (Fig. 6.14).

Escarpment

It is the steep slopes of a hill, a ridge, or a plateau, forming a long barrier rising above the surrounding land. It is formed either by erosion or by faulting. It is represented by the longitudinal contours in which one side has closer contours as compared to other side (Fig. 6.15).

Cliff

It is an upland commonly seen on rocky coast having a very steep or vertical rock face. It is represented by contours which meet at the same point but running outwards to the sea (Fig. 6.16).

Re-entrant

It is a depression on the side of a ring ground or hill. It is represented by the contours projecting towards the higher ground (Fig. 6.17).

Volcanic Cones and Craters

It is a land form with a circular base tapering upwards to a point. It is formed due to the accumulation of exploded material of volcanoes. Crater is a funnel-shaped depression at the top of a volcanic cone. It may be produced either by an explosive eruption or by the collapse of the cone. It is represented by closed contours almost circular in space, whose values are increasing towards the centre except near the opening to show the depression (Fig. 6.18).

Gorge

It is a deep valley with very steep and rocky side walls. It can develop only in the region of hard rocks through the rejuvenation of a stream. It is represented by the contours closer on both sides with decreasing

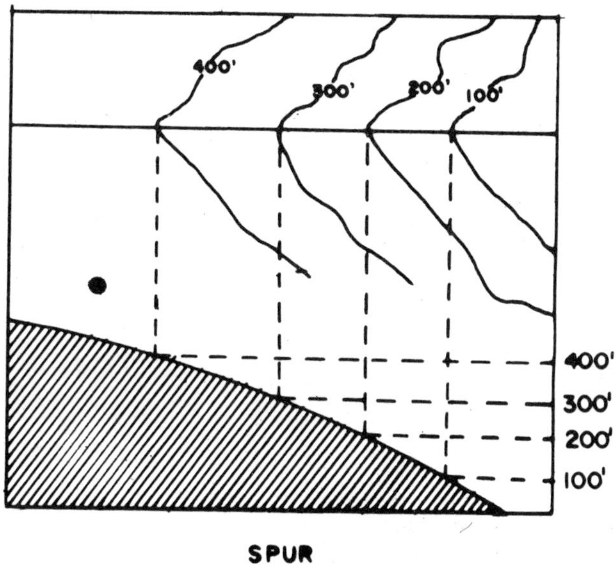

FIG. 6.13: Representation of spur

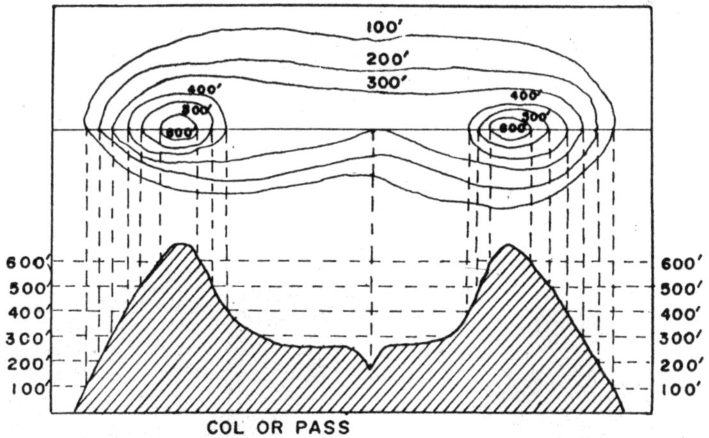

FIG. 6.14: Representation of Col/pass

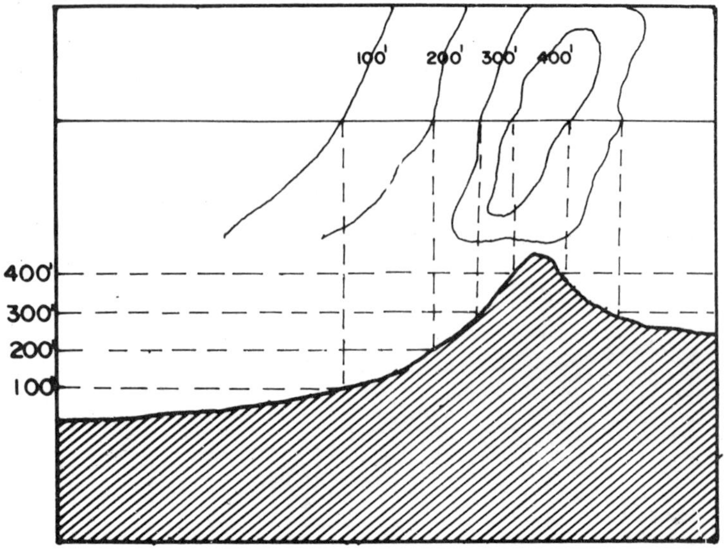

FIG. 6.15: Representation of Escarpment

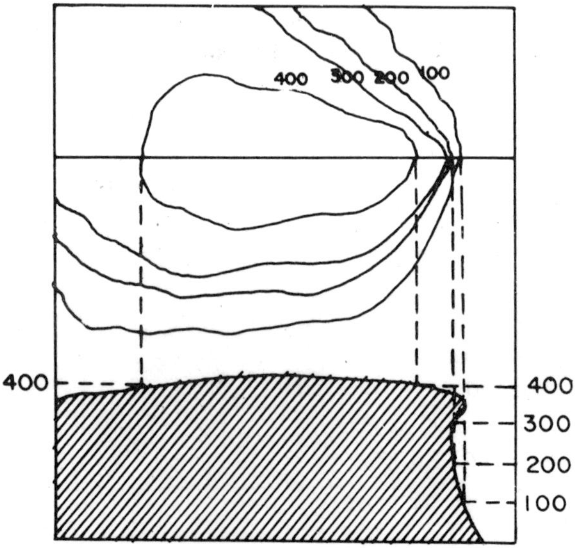

FIG. 6.16: Representation of Cliff

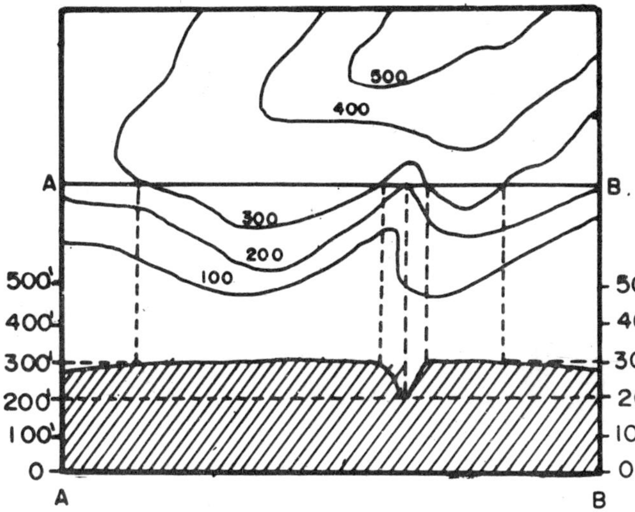

FIG. 6.17: Representation of Re-entrant

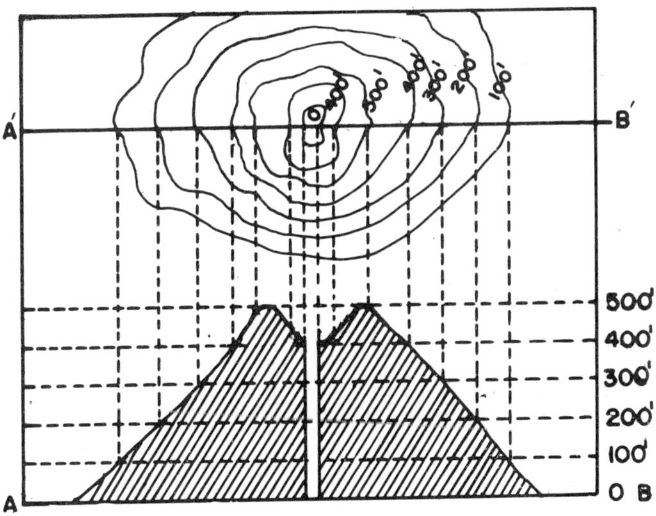

FIG. 6.18: Representation of cone and crater

values towards centre (Fig. 6.19).

V-Shaped Valley

It is generally a narrow valley having the convex slope sides: It is always formed by a river in its young stage. The contours which cut the valley-line are closer together near the head of the valley than further down (Fig. 6.20).

Water-fall

It is a steep cliff-like section of a river channel from which water falls vertically. It is represented by the contours crossing the river and come closer to one another or meet at the same point (Fig 6.21).

Rapid

The bed of the river suddenly becomes steeper and the surface is often broken. A rapid is, therefore, shown by contour lines, which while crossing the valley have come quite close to one another (Fig 6.21).

U-Shaped Valley

The valley is generally formed by glacier erosion. The flatness of the valley bottom is often accentuated by deposits of coarse sediments. The valley has a flat floor and steep parallel sides. It is represented by close parallel contours running on two sides of the stream, leaving a broad flat valley floor. (Fig. 6.22)

Cirque

It is a deep amphitheatre-shaped rock basin with steep sides and an opening downstream found in glaciated upland areas. It is represented by the contours which come closer and at their turning point they form almost a circle (Fig 6.22).

Hanging Valley

It is a tributary valley that lies above the main valley and is separated from it by a steep slope. It is represented by contours, which is similar to that of a U-shaped valley on the bank of the main valley (Fig. 6.22).

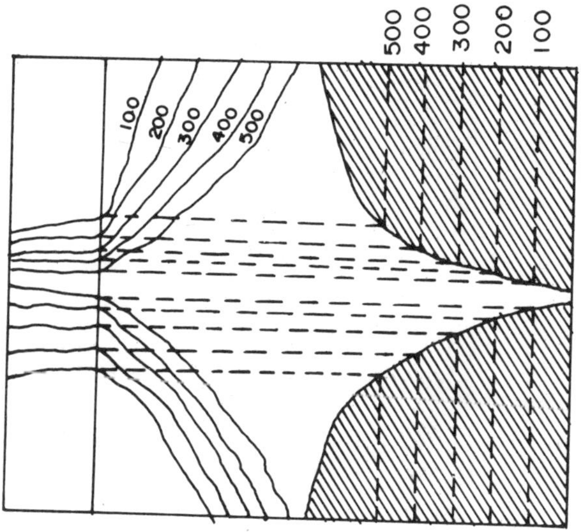

FIG. 6.19: Representation of Gorge

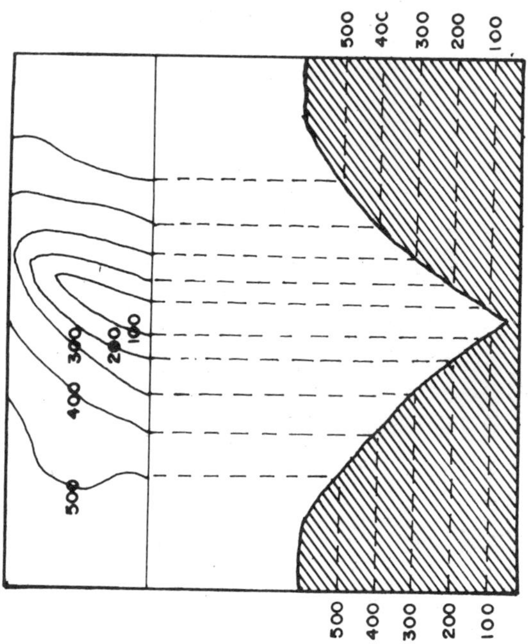

FIG. 6.20: Representation of V-shaped valley

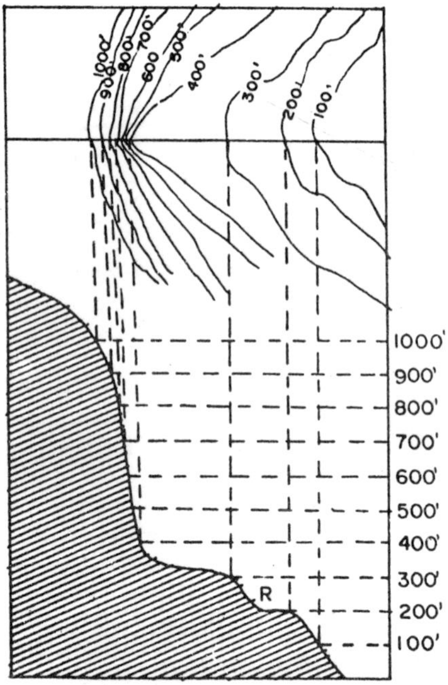

FIG. 6.21: Representation of Waterfall and Rapid

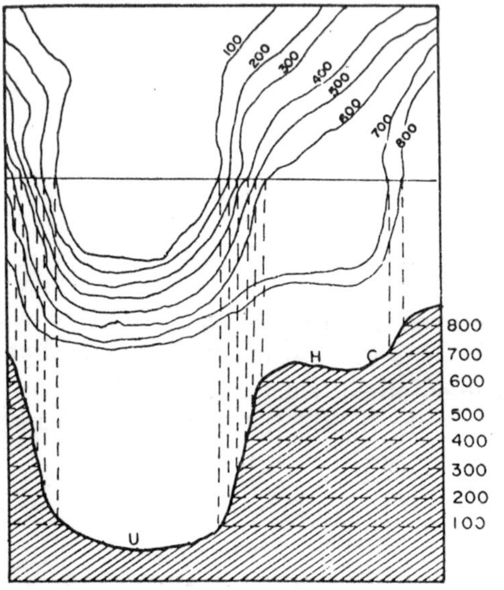

FIG. 6.22: Representation of cirque, hanging valley and U-shaped valley

Ria Coast

A drowned river valley forming a long funnel-shaped inlet of the sea. It has V-shaped cross profiles. Mostly it occurs when the sea-level rises. It is represented by contours, which is closer to the land mass but they are a part of ocean (Fig 6.23).

Fiord Coast

It is formed by glacial action. The glacial valley has been inundated by the sea and forms a deep steep-sided coastal inlet, with a U-shaped cross profile. It is represented by close contours (Fig 6.24).

Sand-Dunes

It is a mound or ridge of sand found in deserts and coastal areas. It is represented by elongated shaped contours (Fig 6.25).

Profiles

A profile is the outline produced where the plane of a section cuts the surface of the ground. It is the easiest method of showing various landforms. In the drawing of an accurate profile, firstly, take a straight edge of paper, put it along the chosen line on the map, then mark accurately all contour intersections, spot-heights, rivers, summits and other defined points. Draw the base-line of profile on a sheet, and transfer the marks carefully. Construct a vertical scale carefully at either end of the base line. and perpendicularly mark the position of each point according to the vertical scale. When all points are plotted, join them by a smooth line. There are four types of profile.

(i) Serial Profile

It is obtained through drawing a number of parallel lines on a relief map in such a way that important features do not remain untouched. Then draw profile along every base line. According to necessity, exaggerate the vertical scale to give prominence to the different relief features.

(ii) Superimposed Profile

Super imposed profiles are those profiles in which all the profiles of series are traced on a single base line. It provides a complex nature of

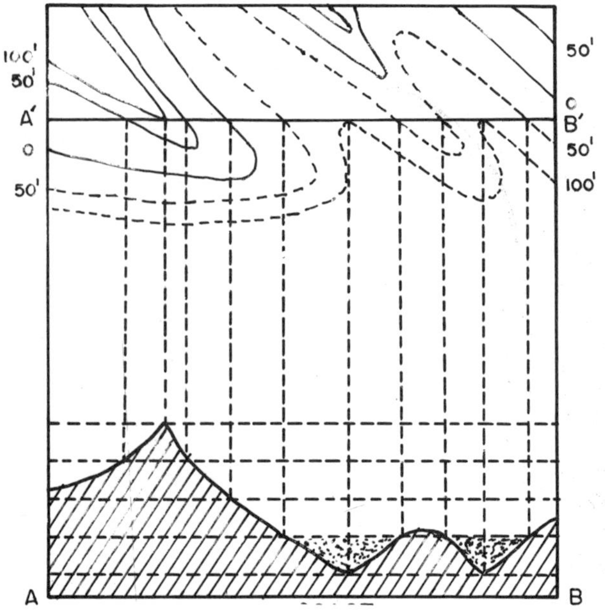

FIG. 6.23: Representation of Ria coast

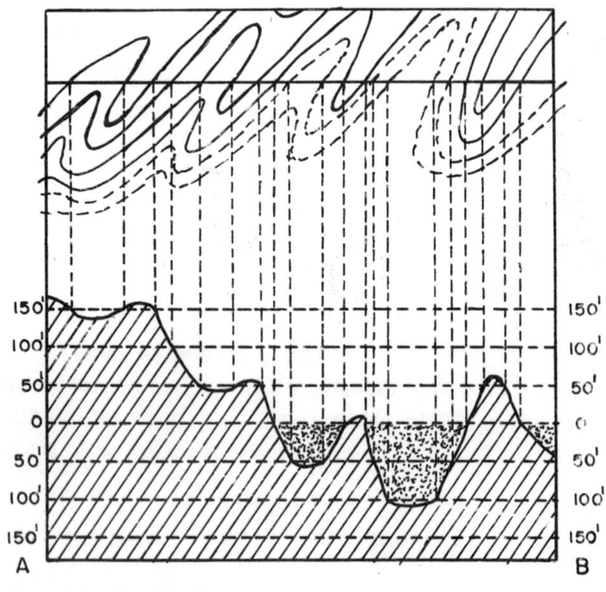

FIG. 6.24: Representation of fiord coast

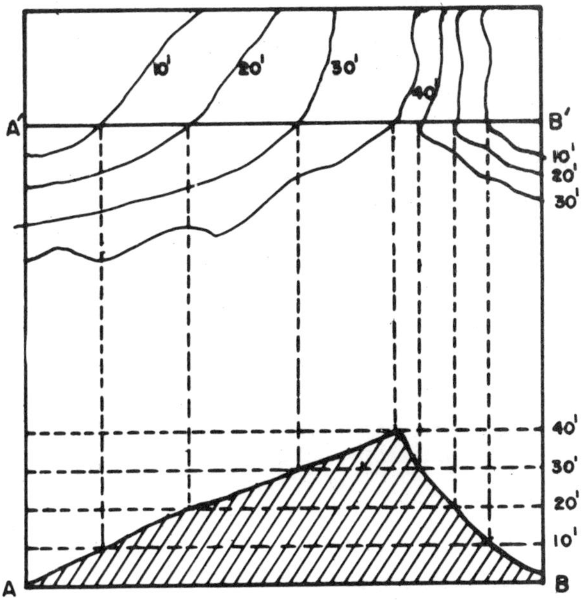

FIG. 6.25: Representation of sand dunes

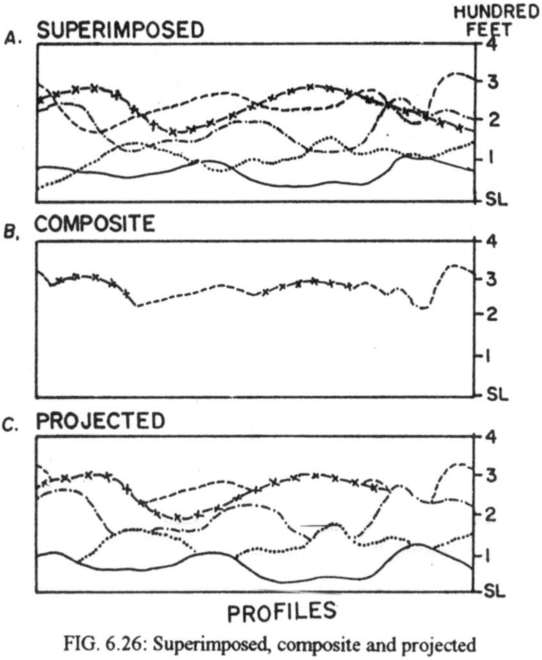

FIG. 6.26: Superimposed, composite and projected

landforms. Each individual line should be carefully numbered for convenience (Fig 6.26A).

(iii) Composite Profile

It is constructed to represent the surface of any area of relief, which include only the highest parts of a series of parallel profiles (Fig 6.26B).

(iv) Projected Profile

It is careful super-imposition of a series of profiles one over the other. In this case, those points of serial profiles are omitted which are below the succeeding profiles. Thus, it provides a panoramic effect, with a distant stay-line, a middle ground and a fore ground (Fig. 6.26C).

7

MAP PROJECTIONS

7.1 DEFINITION

Map projection is the method by which the curved surface of the earth or a part of it, is represented on flat surface on a certain scale by making parallels and meridians. The globe is divided into various sectors by the line of latitude and longitude. The network of these lines is known as graticule. Thus simply map projection is defined as the preparation of the graticule on a flat surface. Mathematical formula are used to construct a graticule on the map corresponding to the intersecting parallels and meridians on the earth.

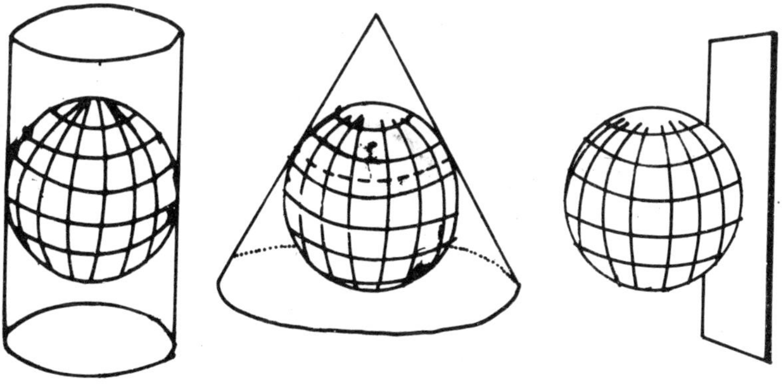

FIG 7.1: Representation of projection

7.2 CLASSIFICATION

Map projection varies according to the size and location of different areas on the earth surface. So in the transformation of spherical surface into a plane surface, following facts should be kept in mind.

(i) Preservation of area,
(ii) Preservation of shape,
(iii) Preservation of bearing (direction and distance).

It is very difficult work to make a projection in which all the three qualities can be well preserved easily. On the basis of these dificulties, the following groups of projections have been prepared.

7.3 CONSTRUCTION OF MAP PROJECTION

1. Equal area or homolographic projection
2. Correct shape or orthomorphic projection
3. True bearing or azimuthal projection

In the first group of projection the graticule is prepared in such a way that every quadrilateral of graticule may appear proportionately to the corresponding spherical quadrilateral.

The second group is also known as conformal projection. Only few points of the sphere can be easily projected in their true form over a plane surface.To achieve the quality of orthomorphism, the scale is changed from point to point.

In the third group of projections, correct bearing are preserved. The line of sight in every case is normal to the plane of projection at the central point.

There are various ways by which the globe can be projected over a surface. According to the surface used for transferring the graticule, there are three types of projections.

(a) Cylindrical projection—on a cylindrical surface
(b) Conical projection—on a conical surface
(c) Zenithal projection—on a plane surface
(d) Conventional projection—it is purely based on mathematical calculation. It is very useful, because, it fulfils our special requirements.

Thus now we have innumerable projections possessing one property or the other. There is no projection which can be grouped in single class. There can be as many classification, or many bases. Hence following classification may be suggested depending on different bases:

A. Based on the method of construction

1. Perspective
2. Non-perspective

B. Based on the developable surface used

1. Conical
2. Cylindrical
3. Zenithal
4. Conventional

C. Based on the preserved quantities

1. Homolographic or equal area projections
2. Orthomorphic
3. Azimuthal or true bearing projections

D. Based on the position of tangent surfaces

1. Polar
2. Equatorial or Normal
3. Oblique

E. Based on the position of view point or light

1. Gnomonic
2. Stereographic
3. Orthographic

F. Based on geometric shape

1. Rectangular
2. Circular
3. Eliptical
4. Butterfly shape

(i) Natural Cylindrical Projection

This projection is also known as perspective or natural cylindrical projection. It is constructed as if a rectangle of paper upon which the details of the earth's surface are projected, formed into a cylinder, surrounds the globe. The paper is then rolled out flat. Meridians and parallels appear as straight lines. On this type of projection the polar regions cannot be shown accurately and territories near the poles are exaggerated in size.

Properties

(1) All the parallels and meridians are straight and meet one another at right angles;
(2) The meridians are equi-spaced but the parallels are not equi-spaced. Although parallels are equal to the equator in length;
(3) East-West and North-South scales are exaggerated. The amount of exaggeration is not the same in both the direction and hence the area is increased and shape is distorted; and
(4) Distances are correct near the equator only.

Mathematical Construction

Calculate the length of equator $2\pi R$, where R is the radius of the earth reduced to scale and length of an interval along the equator equal to

$$\frac{2\pi R \times \text{interval}}{360}$$

Example : Draw a Natural cylindrical projection on the scale 1:312,500,000 interval being 15°

Radius (R) = 250,000,000/312,500,000 = 0.8 inch

$$
\begin{aligned}
\text{Length of the equator} &= 2\pi R \\
&= 2 \times 22/7 \times 0.8 \\
&= 35.2/7 \\
&= 5.02'' \\
\text{Interval along the equator} &= 2\pi R \times \text{interval}/360 \\
&= \frac{2 \times 22 \times 0.8 \times 15}{7 \times 360} \\
&= 5.02 \times 15/360 \\
&= 75.3/360 \\
&= 0.2''
\end{aligned}
$$

Draw a half circle with the radius of 0.8" and mark the interval of 15° in Northern and Southern Hemisphere. Extend *AB* upto *C*, which represents the equator 5.02". Draw perpendicular line at *B* and *C* to equator . Extend the tangent line upto the perpendicular line. Draw the parallel line to equator *BC* from the cutting point of tangent and vertical lines. Mark the interval along the equator at 0.2" of 15° (Fig. 7.2).

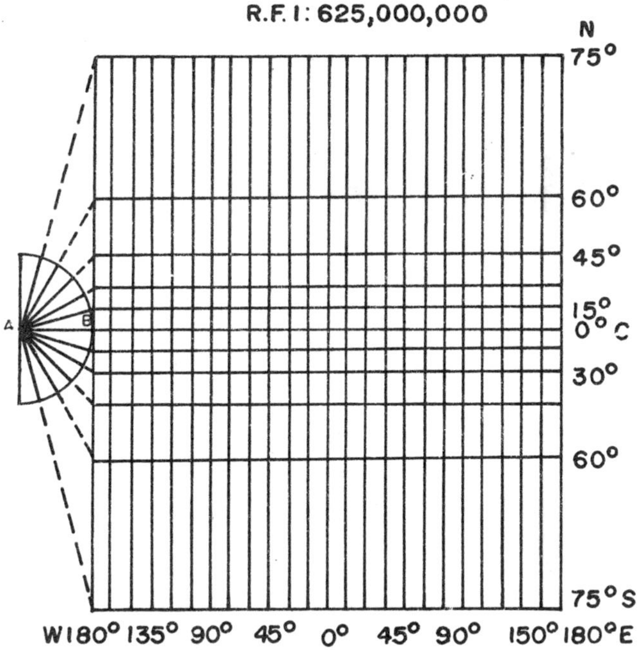

FIG. 7.2: Natural cylindrical projection

(ii) Simple Cylindrical Projection

It is also known as the Plate Carree or equi-distance projection because the meridians and parallels are drawn at equal distance.

Properties

1. Parallels and meridians are straight line;
2. Parallels and meridians are equi-spaced and intersect at right angles;

3. The scale along the equator is correct to show the correct distance along the equator but all other parallels are exaggerated;
4. The meridians are drawn true to the scale; and
5. The projection is neither equal—area nor orthomorphic.

Uses : There are little uses of this projection because it shows correctly upto 2° 30' N and S of the equator.

Example : Draw a simple Cylindrical projection on the scale 1:416666670 at 15° interval.

Radius *(R)*	= 250,000,000/416,666,670 = 0.6"
Length of the equator	= 2 × 22/7 × 0.6
	= 26.4/7
	= 3.77"
Interval along the equator	= $2\pi R$ × interval/360
	= 3.77 × 15/360
	= 56.55/360
	= 0.15"

Draw a half circle of 0.6" radius and mark the 15° interval. Extend the *AB* upto *D* which represents the equator of 3.77". Mark the number of intervals along the equator at 0.15" which is equal to the arc *BD* of angle 15° . Draw perpendiculars to the equator from all these points and mark intervals of 0.15" along any one of them. Now, draw the parallel line to the equator from the points which is along the perpendicular line and complete the projection (Fig. 7.3).

(iii) Cylindrical Equal Area Projection

A projection in which meridians are shown parallel lines, spaced at their true to scale distance from the equator Lines of latitude are also parallel but are closer near to the equator and wider apart towards high latitudes. As a result the map has equal area but is highly distorted towards the poles.

Properties

1. The projection is an equal area projection;
2. The parallels and meridians are all straight lines meeting at

right angles to each other;

3. The distance between the parallels decrease towards the north and south poles; and
4. The shape of the map is distorted at higher latitude and the area represented becomes incorrect.

Uses: Being an equal-area projection, it is used for showing distributional pattern suitably upto 45° N and S of the equator.

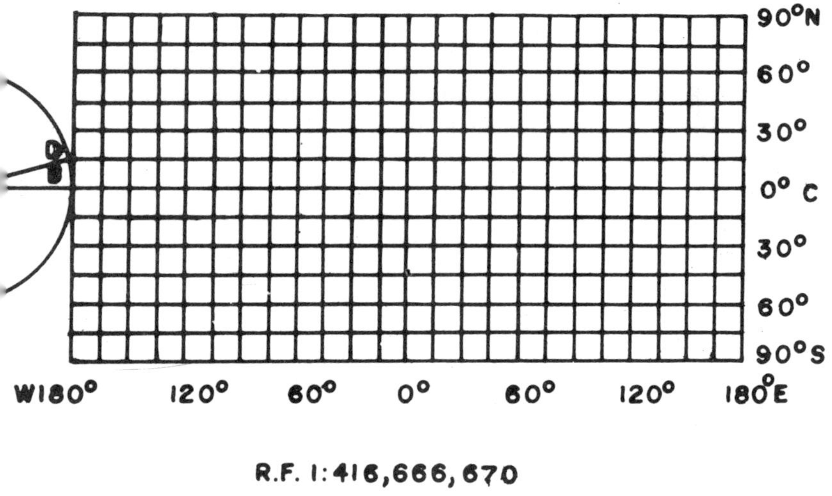

FIG. 7.3: Simple cylindrical projection

Example : Draw a cylindrical equal area projection for the world map on the scale of 1:360,000,000 at 15° interval.

Radius (R)	= 250,000,000/360,000,000
	= 0.69 = 0.7"
Length of the equator	= 2π R
	= 2 × 22/7 × 0.7
	= 30.8/7
	= 4.4"
Interval along the equator	= $2\pi R$ × interval/360

$= 4.4 \times 15/360$
$= 66/360$
$= 0.18"$

Draw a half circle with the radius of 0.7" and parallels at the interval of 15° . Extend the *AB* upto *C*, which represents the equator 4.4" . Draw the parallel line to the equator *(BC)* from the meeting point of radius and circumference of the half circle. Mark the interval along the equator at 0.18" of 15° and draw perpendicular lines to the equator from all these points (Fig. 7.4).

(iv) Mercator's Projection

It is also known as Cylindrical Orthomorphic projection. In this type of projection the equator is drawn true to scale but all other parallels are drawn the same length to the equator, therefore, exaggeration increase away from the equator. Scale from a given point is the same in all direction and shape over small areas shown correctly. It is invented by Gerardus Mercator, a Dutch cartographer who first used the projection in 1569. Thirty years later a British cartographer, Edward Wright, modified it.

Properties

1. The parallels and meridians are represented by straight line and meet at right angles to one another;
2. The meridians are equi-spaced but the parallels are not equi-spaced. The distance between parallels increase as one moves away from equator;
3. The direction between two points as well as shape of small areas is correct;
4. Areas are much exaggerated, although over a small area the shape remains correct; and
5. The distances are correctly shown only along the equator.

Uses : It is used to mark the sailing route by means of great circles and Loxodromes. It is also used to show global patterns of ocean currents and wind systems in atlases.

Mathematical Construction

To find the distance of a particular latitude from the equator, use the

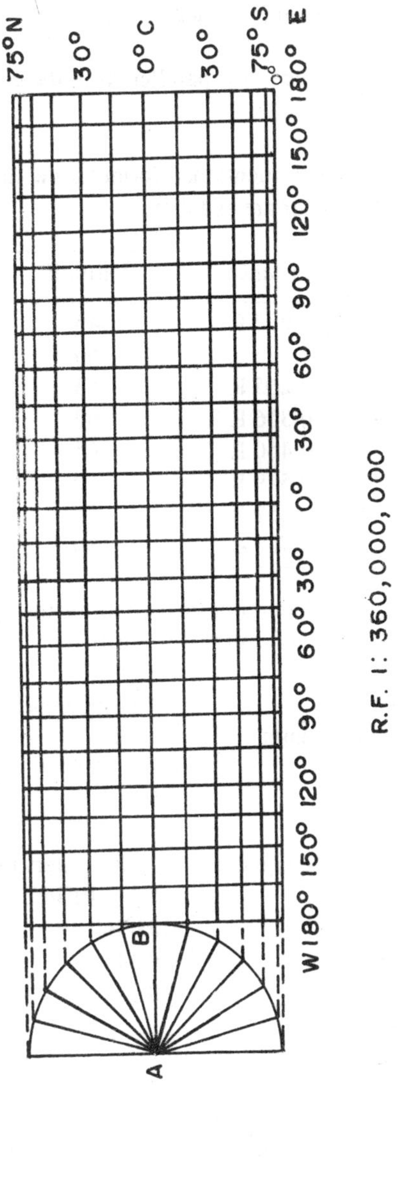

FIG. 7.4: Cylindrical Equal-area projection

following formulae.

$$X = 2.3026\ R \log \tan (45 + \phi/2)$$

where : X is the distance of a parallel from the equator and f is the latitude.

By using the above mentioned formula, the distances of various parallels from the equater are as follow:

Latitude	Distance	Latitude	Distance
5	0.087 R	45	0.880 R
10	0.175 R	50	1.153 R
15	0.265 R	55	1.317 R
20	0.356 R	60	1.505 R
25	0.450 R	65	1.736 R
30	0.549 R	70	2.025 R
35	0.652 R	75	2.437 R
40	0.763 R	80	3.132 R

Example : Draw a Mercator's Projection for the world on a scale of 1:300,000,000 interval being 10°.

Radius (R) = 250,000,000/300,000,000 = 0.83"

Length of the equator = 2π R
= 2 × 22/7 × 0.83
= 5.217
= 5.2"

Interval along the equator = 2π R × interval/360
= 5.2 × 10/360
= 52/360
= 0.144
= 0.14"

Parallels distance from equator are

10 = 0.175 × 0.83 = 0.145 = 0.15"
20 = 0.356 × 0.83 = 0.295 = 0.30"
30 = 0.549 × 0.83 = 0.455 = 0.46"
40 = 0.736 × 0.83 = 0.610 = 0.61"

$$50 = 1.153 \times 0.83 = 0.957 = 0.96''$$
$$60 = 1.317 \times 0.83 = 1.093 = 1.09''$$
$$70 = 1.735 \times 0.83 = 1.440 = 1.44''$$
$$80 = 2.436 \times 0.83 = 2.021 = 2.02''$$

Take the equator *(BC)* equal to 5.2" length. Divide the equator into 36 equal parts of 0.14" each. Draw the perpendiculars from these points of division to get meridians. Mark above mentioned distance from the equator along the central meridian. Draw parallel lines to the equator *(BC)* from these points to complete the projection (Fig. 7.5).

Conical Projection

It is also known as Conic Projections. This type of projection constructed as if the meridian and parallels are drawn on a paper cone capping the globe, which then unrolled. Normally the cone touches the globe along one of the parallels and apex of the cone is above one of the poles through which the prolonged axis of the globe passes. The projection is accurate along the parallel that the cone touches but away from it there is distortion which is considerable. It is, therefore, only used for relatively small sections of the earth surface. Distances can be made accurate if the cone is made to touch the globe along two standard parallels, one near the top and one near the bottom of the map. This projection is used for maps showing considerable EW extent of the earth surface.

The length of the standard parallel = 2n R cos ϕ where R is the radius of the earth reduced to the scale and is the latitudinal value.

(v) Simple Conical Projection with One Standard Parallel

In this type of conioprojection there is only one standard parallel which is true to scale. The standard parallel gets the central position. All the other parallels are drawn parallel to standard parallel from a common centre at equal intervals along the central meridian. Intervals along the standard parallel are found out and joined straigt with the common centre to show meridians.

Properties

1. Parallels and meridians are equi-spaced;
2. Parallels are arcs of concentric circles and meridians are

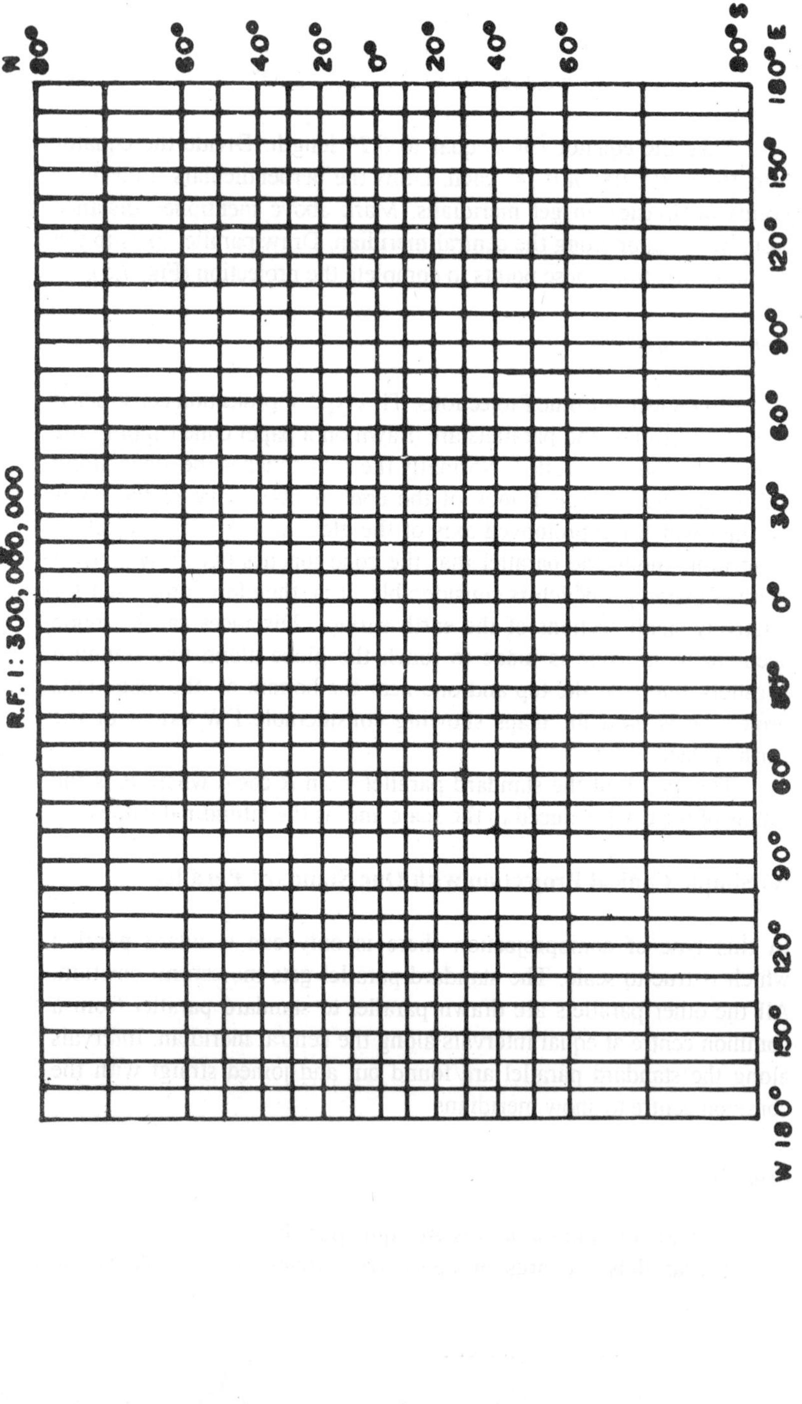

FIG. 7.5: Mercator's Projection

straight lines radiating out from common centre;

3. Pole is represented by an arc of a circle and is different from the centre of the parallels.
4. Scale is true only along the standard parallel; and
5. Both shape and area are correct only along the standard parallel but they are exaggerated away from the standard parallel.

Uses : This projection is used for an area which extends East-West with narrow North-South extensions. Due to exaggeration, the projection is not suitable for large latitudinal extent.

Example : Draw a Simple conical projection with one standard parallel for an area bounded by 40° N and 60° N latitudes and 45° W and 65° W longitudes on the scale of 1:200,000,000 at the interval of 5°.

Standard Parallel is 50° N (40°, 45°, 50°, 55°, 60°)
Central Meridian is 55° W (45°, 50°, 55°, 60°, 65°)

Radius *(R)* =250,000,000/200,000,000
= 1.25"

Interval along the central meridian = 2π R × interval/360
= 2 × 22/7 × 1.25 × 5/360
= 275/7 × 360
= 275/2520
= 0.109
= 0.11"

Interval along the standard parallel = 2π R cosϕ × 5/360
= 2 × 22/7 × 1.25 × cos 50° × 5/360
= 275/7 × .6427/360
= 275/7 × .643/360
= 176.825/7 × 360
= 176.83/2520
= 0.07"

Radius of standard parallel = R cotϕ
= 1.25 × cot 50°
= 1.25 × 0.839
= 1.048
= 1.05"

With these calculations the projection can be drawn as in the graphical case (Fig. 7.6).

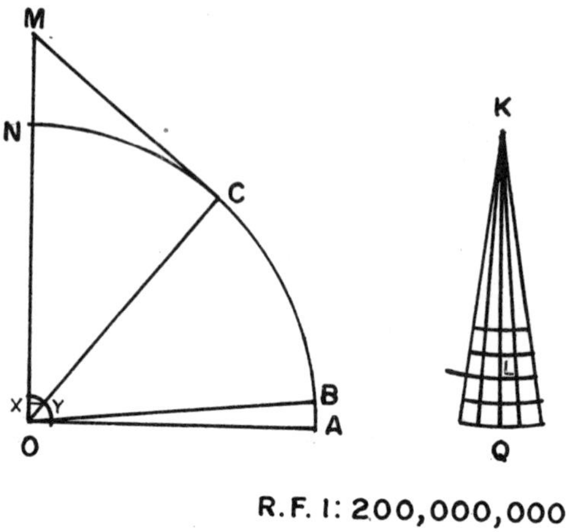

FIG. 7.6: Simple conical projection

Graphical Construction

Draw a quadrant with the radius *(AO)* of 1.25" in length. Draw two angles at the centre *(O)*, and angle $\angle AOB$ equal to the interval i.e 5 and the angle $\angle AOC$ of standard parallel i.e. 50°. Draw a tangent at *C* and extend the axis *(ON)* upto *M* to meet the tangent *(CM)*. Then *CM* is the radius of the standard parallel. The distance of *AB* is the interval along the central meridian. Draw an arc at the centre *(O)* with the radius having distance equal to *AB*. Draw a perpendicular *XY* from *Y* to *ON*. Then the distance of *XY* is the interval along the standard parallel. Now draw a vertical line separately. Draw an arc with the radius equal to the distance of *CM* from point *K* on the vertical line to represent the standard parallel 50° N. The vertical is the central meridian. From the point of intersection of the standard parallel and the central meridian *(L)* mark intervals along the standard parallel equal to *XY* and along the central meridian equal to *AB*. From point *K* draw arcs through the various points marked along the central meridian. Finally join the centre *K* with the points along the standard parallel by means of straight lines and complete the projection (Fig. 7.6).

(vi) Conical Projection with Two Standard Parallel

In this type of conic projection there are two standard parallels which are true to scale. The areas can be represented more correctly along these two standard parallels and as one moves away from these parallels the amount of errors increases. In any representation, therefore, the selection of these two standard parallels is very important. They are generally selected in a manner in which two-thirds of the area between these and one-sixth of the area remains outside each standard parallel.

Properties

1. Parallels are arcs of circles and are equi-spaced;
2. Meridians are straight lines radiating from common centre;
3. Pole is represented by an arc of a circle;
4. Scale is true along the two standard parallels but exaggerated away from the standard parallel;
5. Scale is also true along the central meridian by construction as well as other meridians because other meridians are the radii of the same circle similar to the central meridian; and
6. Representation of distances between the standard parallels are shorter than the actual distances but outside the standard parallels representation of distances are greater than the actual distances.

Uses : The use of this projection is suitable in the middle latitudes for representing an area which extends East-West with considerable North-South extensions. Because of two standard parallels it is used for larger countries e.g. Canada.

Example : Draw a conical projection with two standard parallels for an area bounded by 25° N to 75° N latitudes and 20° E to 100° E longitudes on the scale of 1: 160,000,000 with an interval of 10°.

Standard Parallels are 35° N & 65° N (25° , 35° , 45° , 55° , 65° , 75°)
Standard Meridian is 60° E (20°, 30°, 40°, 50°, 60°, 70°, 80°,90°,100°)

Radius (R) = 250,000,000/160,000,000
= 1.56"

Interval along the central meridian

$= 2p\,R \times \text{interval}/360$
$= 2 \times 22/7 \times 1.56 \times 10/360$
$= 686.4/2520$
$= 0.27''$

Interval along the 35th parallel

$= 2\pi R \times \cos 35^o \times \text{interval}/360$
$= 2 \times 22/7 \times 1.56 \times 0.819 \times 10/360$
$= 56.21/7 \times 36$
$= 56.21/252$
$= 0.22''$

Interval along the 65th parallel

$= 2\pi R \times \cos 65^o \times \text{interval}/360$
$= 2 \times 22/7 \times 1.56 \times 0.422 \times 10/360$
$= 28.96/252$
$= 0.11''$

Radius of the 65th parallel *(X)*

$= [2\pi R \;(X—y/360) \times (\cos X/\cos y - \cos X)]$
$= [2 \times 22/7 \times 1.56\;(65\text{-}35/360) \times (\cos 65/\cos 35\text{-}\cos 65)]$
$= 68.64/7 \times 30/360 \times (0.422/0.819\text{-}0.422)$
$= 205.92/252 \times 0.422/0.397$
$= 86.89/100.04$
$= 0.87''$

Radius of the 35th parallel *(y)*

$= 0.87''$ inch $+ 2\pi R\;(\times\text{-}y/360)$
$= 0.87 + 2 \times 22/7 \times 1.56\;(65\text{-}35/360)$
$= 0.87 + 68.64/7 \times 30/360$
$= 0.87 + 68.64/7 \times 1/12$
$= 0.87 + 68.64/84$
$= 0.87 + 0.81$
$= 1.68''$

With these calculation the projection can be drawn as in the case of graphical construction (Fig. 7.7).

Graphical Construction

Draw a quadrant with the radius *(AO)* of 1.56" in length. Draw three angles at the centre *(O)*, an angle $\angle AOB$ equal to the interval and the two angles $\angle AOC$ and $\angle AOD$ of the two standard parallels. Draw another quadrant at the centre *'O'* with the radius having the distance equal to *AB*. The arcs meets *OC* and *OD* in *'x'* and *'y'* . From *'x'* and *'y'* draw perpendiculars *xx'* and *yy'* on the axis *OE*. *XX'* and *YY'* are intervals along 35° and 65° standard parallels respectively. Now draw a vertical line and mark intervals equal to *AB* along it. At standard parallels (35° and 65°) draw perpendiculars *MP* and *QR* equal to *XX'* and *YY'* respectively. Join *PR* and extend it to meet the vertical line in *'n'*. Then *NQ* and *MN* are the radii of 65° and 35° standard line *EF*, which represent central meridian an dmark the radii *NQ* and *MN* along it. Mark intervals along the central meridian equal to *AB* between *FG* and outside *FG* according to the requirements. Draw standard parallels as well as other parallels with *E* as centre which passes through the points marked along the central meridian. Mark intervals along the standard parallels equal to *XX'* and *YY'* respectively. Join these points along the standard parallels with centre *E* and complete the projection (Fig. 7.7).

(vii) Bonn's Projection

It is a modified conical projection in which exaggerations of scale are removed along the prallels in the sense that all parallels are true to scale. The central meridian is straight, while the others are drawn by joining the intervals along the parallels by smooth curves, The parallels are concentric circles. It is an equal area projection. It is easy to draw with reasonable shapes around the central meridian.

Properties

1. All the parallels are true to scale and are concentric curves;
2. Central meridian is a straight line and other meridians are smooth curves;
3. The scale is correct only along the central meridian; along the other meridians there is exaggeration which is increases away from the central meridian. Due to that the shape becomes increasingly distorted away from the central meridian; and
4. It is an Equal area projection.

Uses : This projection is used for showing large countries and continents. Asia is not very well shown because there is little distortion in the shape of the margins. Although in atlas this projection has been used for all continents except Africa. It has been also used for topo-sheets by a number of countries in Europe.

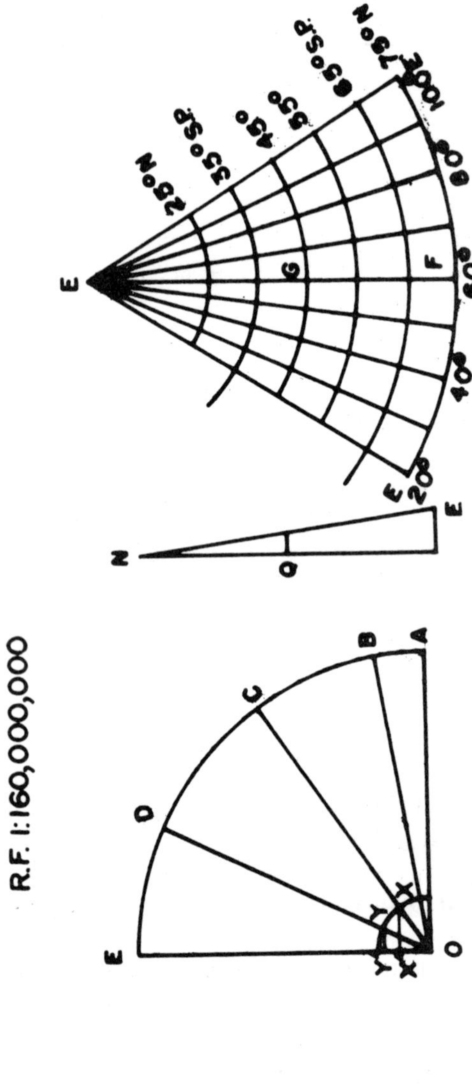

FIG. 7.7: Conical Projection with two Standard Parallels

Example : Draw a Bonn's Projection for an area bounded by 25° N to 65° N latitude and 30° W to 30° E longitude on a scale 1: 200,000,000 and interval 10°.

Central Parallel is 45° N (25° , 35° , 45° , 55° , 65°)
Central Meridian is 0° (30° , 20° , 10° , 0° , 10° , 20° , 30°)

Radius *(R)* = 250,000,000/200,000,000= 1.25"

Radius of the 45° N parallel = R cot Lat.
= 1.25 × cot 45°
= 1.25 × 1
= 1.25"

Interval along the central meridian (C.M) = $2\pi R$ interval/360
= 2 × 22/7 × 10/360
= 44/252
= 0.17
= 0.2"

Interval along the 25th parallel = Interval along C.M × cos 25
= 0.22 × cos 25
= 0.22 × 0.90 = 0.20"

" " 35th " = 0.22 × cos 35
= 0.22 × 0.81 = 0.17"

" " 45th " = 0.22 × cos 45
= 0.22 × 0.70 = 0.15"

" " 55th " = 0.22 × cos 55
= 0.22 × 0.57 = 0.12"

" " 65th " = 0.22 × cos 65
= 0.22 × 0.42 = 0.09"

With these calculation the projection can be drawn as in the case of graphical construction (Fig. 7.8).

Graphical Construction

Draw a quadrant of a circle with radius *(AO)* of 1.25" in length. Draw angles 10° , 25° , 35° , 45° , 55° , and 65° , at the centre *(O)* to represent the quadrant at the centre *(O)* with the radius having distance equal to *AB.* The arc meets *OC, OD, OE, OP* and *OG* in *L, M, N , P* & *Q* respectively. From *L, M, N, P* and *Q* draw perpendiculars *LL', MM', NN', PP',* and *QQ'* on the axis *OH.* Draw a tangent at *'E'* and extend the axis *(OH)* upto *'I'* to meet the tangent *(EI).* Then *EI* is the radius of the central parallel. The distance of *AB* is the interval along the central meridian. Now draw a vertical line *XY* which represents central meridian and from point *X* draw an arc with the radius equal to the distance of *EI* to represent the central meridian. Mark other intervals equal to the distance of *AB* and through these points draw arcs of circles with *X* as centre. Mark intervals along 25° ,35° , 45° , 55° , and 65° , degree parallels equal to the distance of *LL', MM', NN', PP'* and *QQ* respectively. Now join the points along the parallels by means of smooth curves and complete the projection (Fig. 7.8).

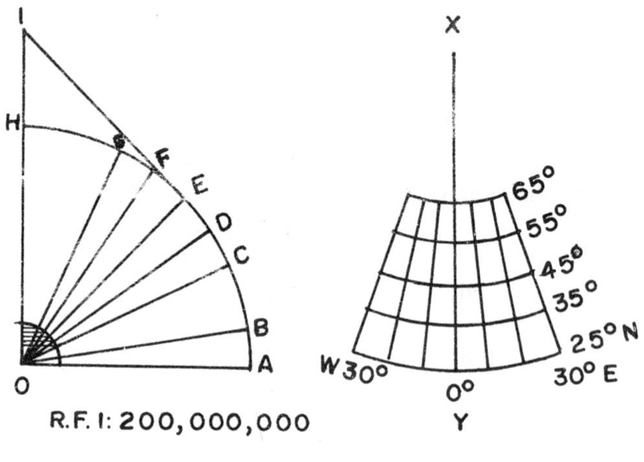

FIG. 7.8: Bonn's projection

(viii) Polyconic Projection

It is a modified form of conical projection in which cutting point of all parallels to the central meeridian is true to scale. The distances between the meridians along each parallel are also true to scale. Near the central meridian both area and shape are approximately correct, but away from it both are wrong.

Properties

1. Parallels are arcs, having different cnetre;
2. Except the central meridian, all other meridians are regular curves. The central meridian is a straight line;

3. All parallels are standard parallels;
4. East west distances are correctly represented along the particular parallels and north south distances are correctly shown only along the central meridians;
5. The projection is neither equal area nor orthomorphic.

Uses : The projection is suitable for representing areas on small sheets and such sheets can fit together on their northern and southern ends. This projection is suitable for representing small areas. Areas with a large latitudinal and limited longitudinal extent in the middle latitudes can be represented cn this projection.

Example : Draw a polyconic projection for an area bounded by 25° N to 65° N latitude and 20° W to 60° E on the scale 1: 150,000,000 and interval 10°.

Central meridian is 20° E (20°, 10°, 0°, 10°, 20°, 30°, 40°, 50°, 60°)
Central parallel is 45° (25°, 35°, 45°, 55°, 65°)

Reduced earth radius *(R)* = 250,000,000/150,000,000
= 1.67"

Interval along the central meridian (C.M) = $2\pi R$ interval/360°
= 2 × 22/7 × 1.67 × 10/360
= 73.48/7 × 1/36
= 73.48/252
= 0.29"

Interval along 25th parallel = C.M. × cos 25°
= 0.29 × cos 25°
= 0.29 × 0.906 = 0.26

Interval along 35th parallel = 0.29 × cos 35°
= 0.29 × 0.819 = 0.24"

Interval along 45th parallel = 0.29 × cos 45°
= 0.29 × 0.707 = 0.21"

Interval along 55th parallel = 0.29 × cos 55°
= 0.29 × 0.573 = 0.17"

Interval along 65th parallel = 0.29 × cos 65°
= 0.29 × 0.422 = 0.12°

With these calculations projection can be drawn as in the case of graphical construction (Fig. 7.9)

Graphical Construction

Draw a quadrant of a circle with the radius *(AO)* of 1.67" in length. Draw angles 10° , 25° , 35° , 45° , 55° , and 65° at the centre 'o' to represent the interval and other parallels. Draw another quadrant at the centre *'O'* with the radius having distance equal to AB. The arc meets *OC, OD, OE, OF,* and *OG* in *L, M, N, P* and *Q* respectively. From *L, M, N, P* and *Q* draw perpendiculars *LL', MM', NN', PP',* and *QQ'* on the axis *OH.* Draw tangent at *C, D, E, F,* and *G* extend the axis *OH* upto *G'* to meet the tangent *CC', DD', EE', FF',* and *FF'.* Then *CC', DD', EE', FF',* and GG' are the radii of various prallels. Now take a vertical line *'RS'* which represents central meridian (20 E). Mark intervals equal to the distance of *'AB'* from point *'S'* as *S, T,*

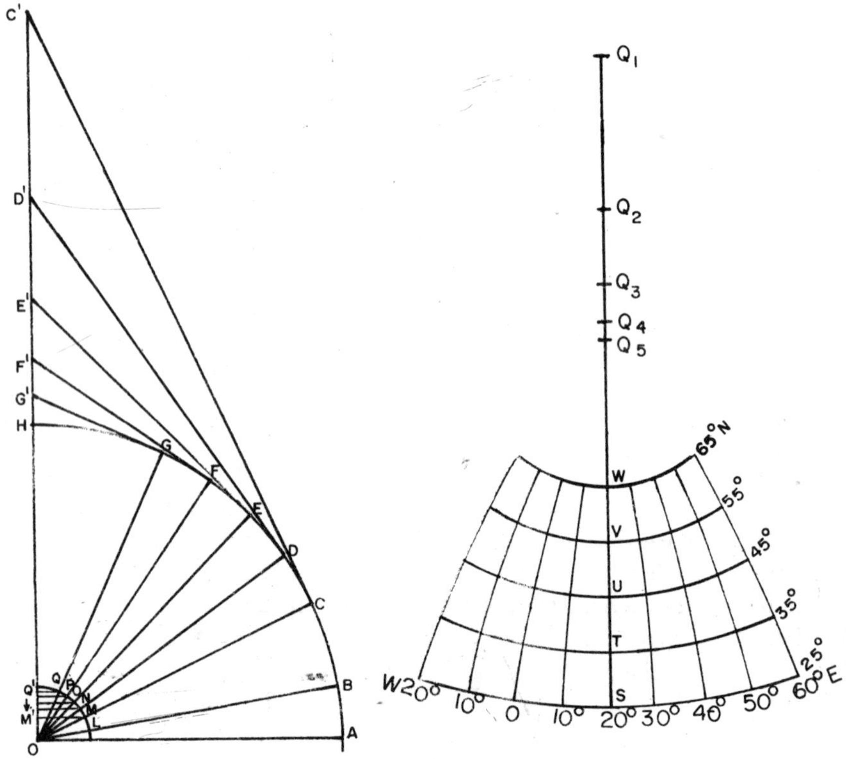

R.F. 1 : 150,000,000

FIG. 7.9: Polyconic projection

U, V, and W. Measure distance equal to *CC', DD', EE',FF'*, and *GG'* from the points *S, T, U, V* and *W* along the line *'RS'* respectively and thus find Q_1, Q_2, Q_3, Q_4, and Q_5, which become the centre of their respective parallel. Now with these centres and corresponding radii draw arcs of circles to show parallels. Mark intervals along various parallels and join the points by smooth curves and complete the projection (Fig. 7.9).

Zenithal Projection

In this type of projection all the bearings are true to compass direction from the centre of the map. By virtue of having the correct direction from the centre of the projection to various other points this projection is also known as Azimuthal projection. In this projection the tangent point becomes the centre of the projection and parallels are concentric circles whereas the meridians are straight lines, radiating from centre. It can be used for mapping any part of the world. It has two broad divisions.

1. Perspective Zenithal projection.
2. Non-perspective Zenithal projection

(ix) Gnomonic Projection

A type of Zenithal map projection in which the projection is obtained by casting the shadow of the lines of latitude and longitude on a tangent plane by pulling the light on the centre of the globe. According to the position of the tangent plane there are three types of the gnomonic projection : Gnomonic Polar Zinithal projection, when the tangent plane touches a pole; Gnomonic Equatorial Zenithal Projection, when the tangent plane touches any point of the equator and Gnomonic Oblique Zenithal Projection, when the tangent plane touches any other points in between equator and pole. The following are the properties of Gnomonic Polar Zenithal projection.

Properties

1. Parallels are concentric curves but not equi-speed & distance between parallels increases rapidly away from the centre.
2. Meridians are straight lines radiating from centre and distances along the meridians increases more rapidly away from the

centre.

3. Scales of parallels and meridians are exaggerated away from the centre hence, shape is also distorted away from the centre.

Uses : The projection is suitable for a small area specially in the polar regions. The area of the 30° latitudinal extent can be suitably shown by this projection.

Example : Draw a Gnomonic Zenithal projection for an area bounded by 30° and 90° latitudes in Northern hemisphere on the scale 1: 250,000,000 and interval 10°.

Radius (R) = 250,000,000/250,000,000 = 1"

Radius for the 80th parallel	$= R \cot 80°$
	$= 1 \times 0.176 = 0.18"$
" " " 70th "	$= R \cot 70°$
	$= 1 \times 0.363 = 0.36"$
" " " 60th "	$= R \cot 60°$
	$= 1 \times 0.577 = 0.58"$
" " " 50th "	$= R \cot 50°$
	$= 1 \times 0.839 = 0.84"$
" " " 40th "	$= R \cot 40°$
	$= 1 \times 1.191 = 1.19"$
" " " 30th "	$= R \cot 30°$
	$= 1 \times 1.732 = 1.73"$

With these radii draw parallels as concentric circles and mark meridians at interval 10° (Fig. 7.10).

Graphical Construction

Draw a quadrant with radius *'MO'* equal to 1 inch in length. *P* is the north pole at which draw tangent *XY* .Now draw the required latitudes 30° , 40° , 50° , 60° , 70° and 80° at the centre 'O' and extend the lines to meet the tangent plane *XY* in *A, B, C, D, E* and *F* because the source of light lying at the centre *'O'*. Thus *PA, PB ,PC, PD, PE,* and *PF* are the projected radii of the corresponding parallels. With these radii and *N* as the centre, draw concentric curves to show parallels. Mark meridian at the centre *(N)* at 10 interval and complete the projection (Fig. 7.10).

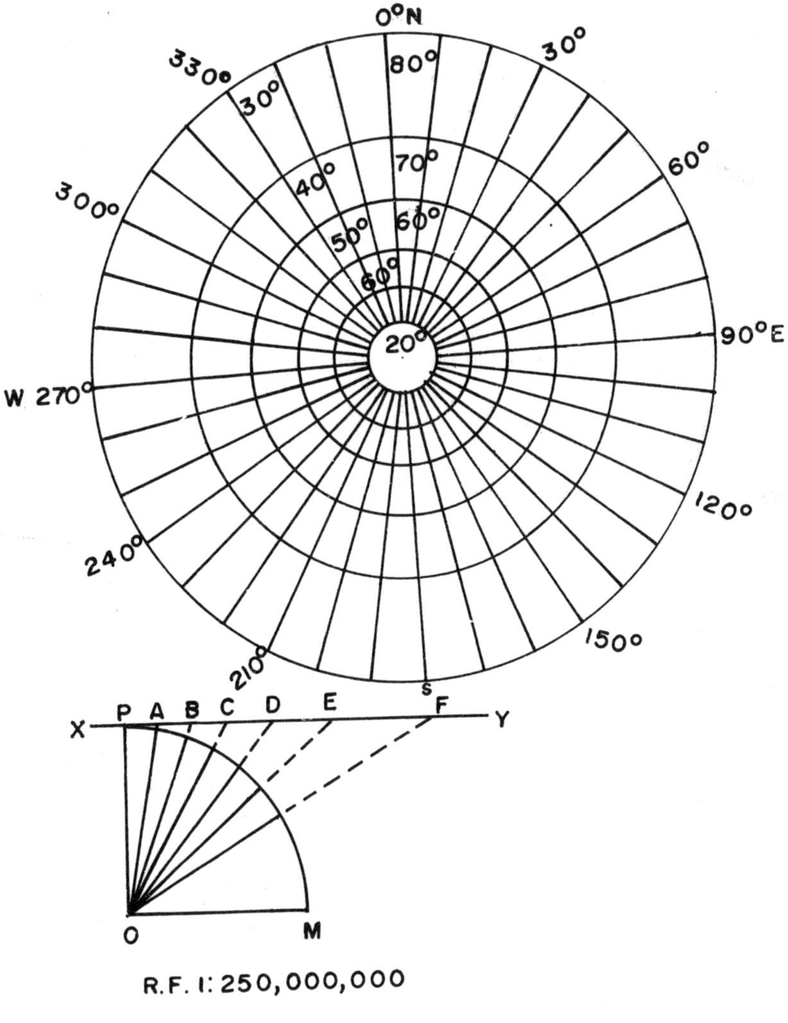

FIG. 7.10: Gnomonic Zenithal projection

(x) Stereographic Projection

This is one of the perspective zenithal projection. This type of map projection has the centre of projection on the circumference of the globe diametrically opposite the position of the sourse of light. According to the position of the tangent plane there are three types of stereographic projection. They are Stereographic Polar Zenithal Projection; Stereographic Equatorial Zenithal Projection and Stereographic Oblique Zenithal Projection. The following are the properties of Stereographic Polar Zenithal projection.

Properties

1. Parallels are concentric curves but not equi-spaced and distances between the parallels increases slowly away from the centre.
2. Meridians are straight lines radiating from the centre and are equi-spaced.
3. Scale along parallels and meridians is exaggerated propoortionately away form the centre, and
4. Shape is maintained correctly.

Uses : This projection is commonly used for the map of the world in hemispheres. It is also suitable for navigational charts for higher latitudes. It is also used to represent the larger countries.

Example: Draw a stereographic Zenithal Projection for the North hemisphere on the scale 1:200,000,000 and interval 10°.

Radius (R)	= 250,000,000/200,000,000
	= 1.25"
Radius for the 80th parallel	= 2 *R* tan (90–80)/2
	= 2 × 1.25 tan 10/2
	= 2.5 tan 5°
	= 2.5 × 0.087 = 0.22"
Radius for the 70th parallel	= 2 × 1.25 tan (90–70)/2
	= 2.5 tan 20/2
	= 2.5 tan 10°
	= 2.5 × 0.176 = 0.44"
Radius for the 60th parallel	= 2 × 1.25 tan (90–60)/2
	= 2.5 tan 30/2

= 2.5 tan 15°
= 2.5 × 0.267 = 0.67"

Radius for the 50th parallel = 2 × 1.25 tan (90–50)/2
= 2.5 tan 40/2
= 2.5 tan 20°
= 2.5 × 0.363 = 0.91"

Radius for the 40th parallel = 2 × 1.25 tan (90–40)/2
= 2.5 tan 50/2
= 2.5 tan 25°
= 2.5 × 0.466 = 1.17"

Radius for the 30th parallel = 2 × 1.25 tan (90–30)/2
= 2.5 tan 60/2
= 2.5 tan 30°
= 2.5 × 0.577 = 1.44"

Radius for the 20th parallel = 2 × 1.25 tan (90–20)/2
= 2.5 tan 70/2
= 2.5 tan 35°
= 2.5 × 0.700 = 1.75"

Radius for the 10th parallel = 2 × 1.25 tan (90–10)/2
= 2.5 tan 80/2
= 2.5 tan 40°
= 2.5 × 0.839 = 2.1"

Radius for the equatorial parallel = 2 × 1.25 tan (90–0)/2
= 2.5 tan 90/2
= 2.5 tan 45°
= 2.5 × 1 = 2.5"

With these radii draw parallels as concentric circles and mark meridian at interval 10 degrees (Fig. 7.11).

Graphical Construction

Draw half circle with radius *MO* equal to 1.25 inches in length. *P* is the north pole at which draw tangant *XY*. Now draw the required latitudes 10, 20, 30, 40, 50, 60, 70 and 80 degree at the centre *'O'*. Throw thelight from the opposite end *'L'*. The ray *'LH'* passes through the point *'H'* which lies at an angular distance of 80 degree from the equator. then *PH* become the radius of 80 degree parallel on the projection because the point *'H'* is projected to *'H'* on the tangant plane. Similarly, 70, 60, 50, 40, 30, 20, 10 and 0 degree parallels are projected to *G', F', E', D', C', B', A'*,and *M* respectively on the

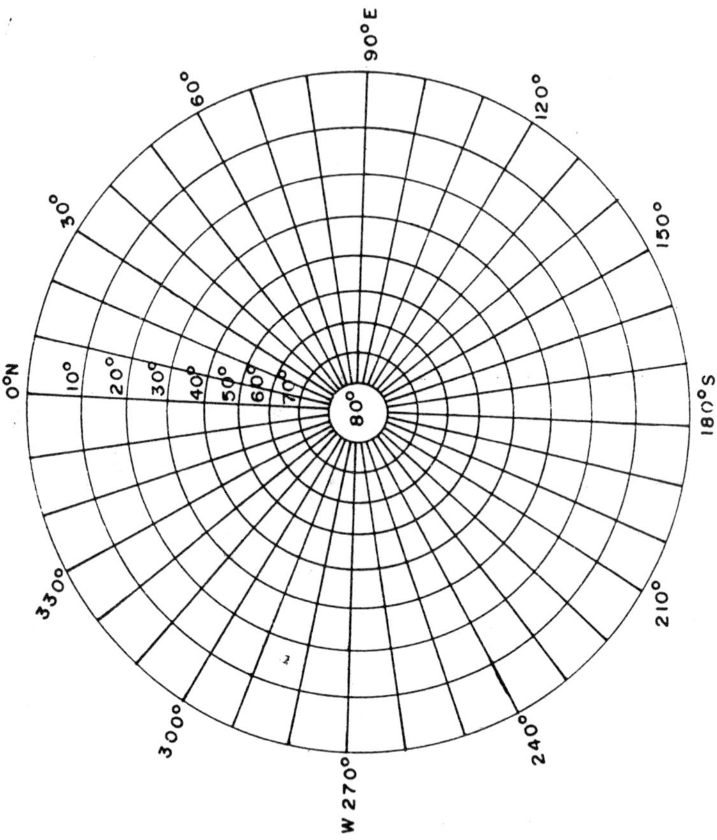

FIG. 7.11: Stereographic projection

tangent plane.Thus, *PG'*, *PF'*, *PE'*, *PD'*, *PC'*, *PB'*, *PA'*, and *PM'*, become the radii of their respective parallels. Now with these radii and *N* as the centre draw concentric curves to show parallels. Mark the meridians at the centre *(N)* at 10 degree interval and complete the projection (Fig. 7.11).

(xi) Orthographic Projection

A map projection in which light is thrown from infinite distance on the tangent plane touching the sphere. The rays of light passing through various latitudes are parallel to each other. According to the position of tangent plane this projection is also sub-divided into three categories i.e. Polar, Equatorial and Oblique. The following are the properties of Orthographic Polar Zenithal Projection.

Properties

1. Parallels are concentric curves but not equi-spaced and distance between the parallels decreases away from the centre,
2. Meridians are straight lines radiating from the centre of the projection,
3. Scale along the parallels is correct but along the meridian scale is not correct.
4. Shape and area are distorted away from the centre.

Uses: The use of this projection is limited specially upto short distance from the centre. Because of the infinite source of light *'d'* is used for astronomical purposes.

Example: Construct a Orthographic Polar Zenithal Projection for the Northern hemisphere on the scale 1:150,000,000 and interval 10°.

Radius *(R)*	= 250,000,000/150,000,000 = 1.67"
Radius for 80th parallel	$= R \cos 80^{\circ}$
	$= 1.67 \times 0.173 = 0.29"$
Radius for 70th parallel	$= R \cos 70^{\circ}$
	$= 1.67 \times 0.342 = 0.57"$
Radius for 60th parallel	$= R \cos 60^{\circ}$
	$= 1.67 \times 0.500 = 0.84"$

Radius for 50th parallel $= R \cos 50^\circ$
$= 1.67 \times 0.642 = 1.07"$

Radius for 40th parallel $= R \cos 40^\circ$
$= 1.67 \times 0.766 = 1.28"$

Radius for 30th parallel $= R \cos 30^\circ$
$= 1.67 \times 0.866 = 1.45"$

Radius for 20th parallel $= R \cos 20^\circ$
$= 1.67 \times 0.939 = 1.57"$

Radius for 10th parallel $= R \cos 10^\circ$
$= 1.67 \times 0.984 = 1.64"$

Radius for 00 parallel $= R \cos 0^\circ$
$= 1.67 \times 1.000 = 1.67"$

With these radii draw parallels as concentric circles and mark meridian at interval 10° (Fig. 7.12).

Graphical Construction

Draw a half circle with radius *MO* equal to 1.67 inches in length. P is the north Pole at which draw tangent *XY*. Now draw required latitude 10°, 20°, 30°, 40°, 50°, 60°, 70°, and 80° at the centre *'O'*. The points *A, B, C, D, E, F, G, H,* and *M* on the surface of the globe are projected as *a, b, c, d, e, f, g, h,* and m through the parallel rays of light *(L)* coming from an infinite distance. With the radii *Pa, Pb, Pc, Pd, Pe, Pf, Pg, Ph* and *Pm* and *N* as centre draw 80°, 70°, 60°, 50°, 40°, 30°, 20°, 10° and 0° Parallels respectively. Mark the meridians at the centre *(N)* at 10° interval and complete the projection (Fig. 7.12).

(xii) Zenithal Equidistant Projection

The parallels are equidistant on the graticule so it is known as equidistant projection. The parallels are projected at their true distances along a meridian. The interval is calculated by the following formula:

$$2\pi R \times \text{Interval of latitude}/360^\circ$$

Properties

1. Parallels are concentric curves and spaced at their true distance apart from one another (equi-spaced).
2. Meridians are straight lines radiating from the centre of the projection.
3. Scale along meridians is correct; and

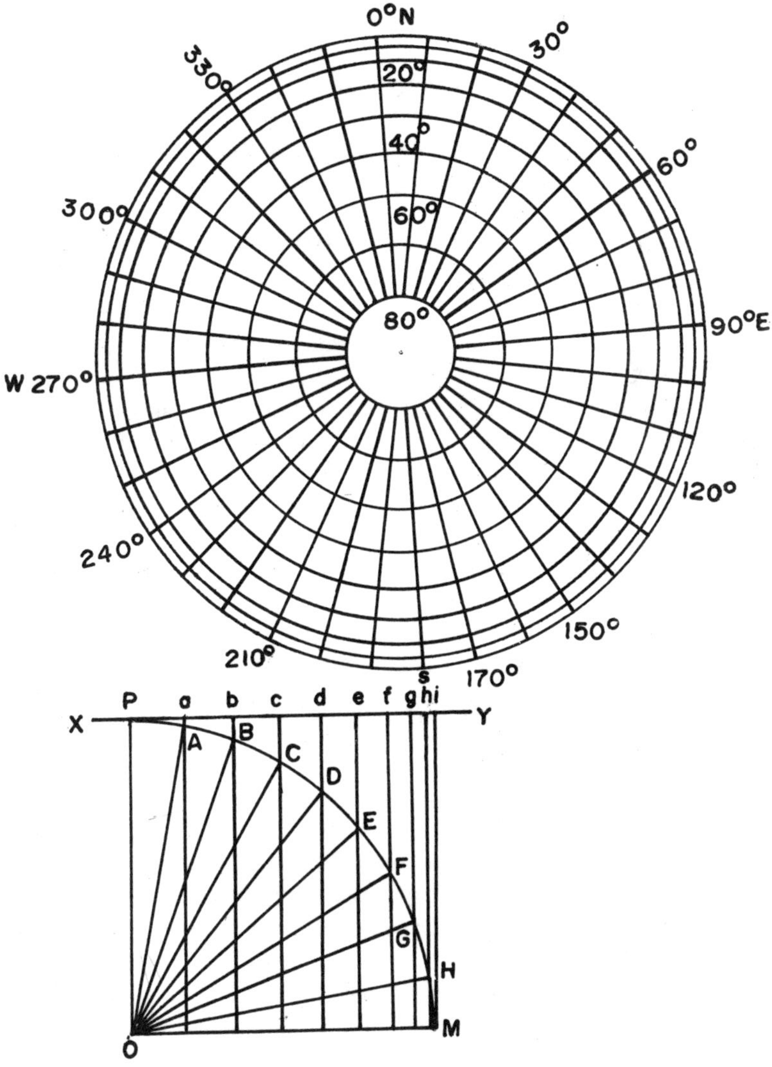

R.F. 1 : 150,000,000

FIG. 7.12: Polar orthographic projection

4. Distances along parallels are exaggerated.

Uses: It is suitable for polar areas extending between 50° to 90°. Occasionally, it is used for hemisphere, but in this case areas near the equator are exaggerated.

Example: Draw a Zenithal Equidistant Projection for the North hemisphere on the scale 1:200,000,000 and interval 10°.

Radius (R)	= 250,000,000/200,000,000 = 1.25"
Length of 10° interval	= $2\pi R$ x interval of lat./360
	= 2 × 22/7 × 1.25 × 10/360
	= 55/252 = 0.22"

Draw a vertical line equal to 1.98" and divide it into 9 equal parts to represent the parallels at 10° interval. With these points draw parallels as concentric circle, with *'O'* as centre. Now mark meridian at 10° interval at the centre with the help of a protractor (Fig. 7.13).

(xiii) Zenithal Equal-Area Projection

This projection was designed by J.H. Lambert in 1772. In this type of projection the parallels have been drawn with different radii so that they enclose on the map the same area as they do on the globe. It is also known as Lambert's Equal area Projection.

Properties

1. Parallals are concentric curves but not equi-spaced and distance between the parallels decreases away from the centre;
2. Meridians are straight lines radiating from the centre of the projection;
3. The area remains correct; and
4. This is an equal-area projection.

Uses: It is frequently uses to represent polar areas, even a hemisphere can be represented correctly. The shape is littlebit distorted away from the centre.

Example: Draw a Zenithal Equal-Area Projection for the North hemisphere on the scale 1:250,000,000 and interval 15°.

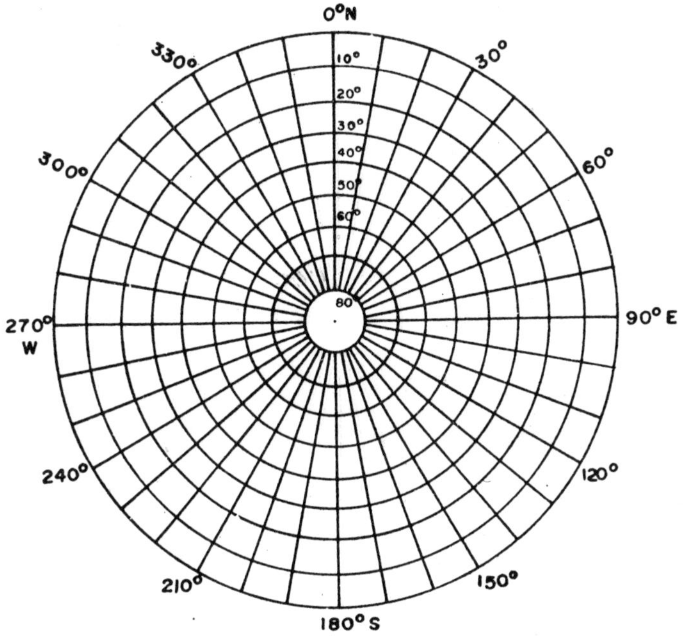

R.F. 1: 200,000,000

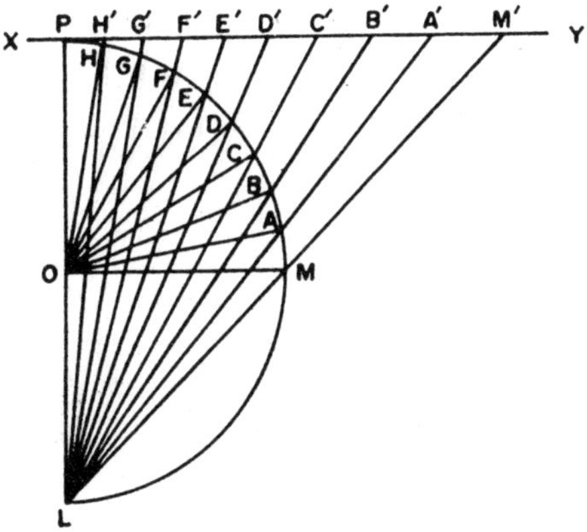

R.F. 1: 200,000,000

FIG. 7.13: Zenithal equidistant projection

Radius *(R)*	= 250,000,000/250,000,000 = 1
Radius for 75th parallel	= 2 R sin (cos – lat)/2
	= 2 × 1 sin (90 – 75)/2
	= 2 sin 15/2
	= 2 sin 7.5°
	= 2 × 0.130 = 0.26"
Radius for 60th parallel	= 2 R sin (90 – 60)/2
	= 2 × 1 sin 30/2
	= 2 sin 15°
	= 2 × 0.258 = 0.52"
Radius for 45th parallel	= 2 R sin (90 – 45)/2
	= 2 × 1 sin 45/2
	= 2 sin 22.5°
	= 2 × 0.382 = 0.76"
Radius for 30th parallel	= 2 R sin (90 – 30)/2
	= 2 × 1 sin 60/2
	= 2 sin 30°
	= 2 × 0.5° = 1"
Radius for 15th parallel	= 2 R sin (90 – 15)/2
	= 2 × 1 sin 75/2
	= 2 sin 37.5°
	= 2 × 0.608 = 1.22"

With these radii draw parallels as concentric circles and mark meridians at interval 15° (Fig. 7.14).

Graphical Construction

Draw a quadrant with radius *'MO'* equal to 1" in length. Now draw the required latitude, (at 15° interval) 15°, 30°, 45°, 60°, 75° and 90° at centre *'O'*, the radii making these angles meet the circumference in *A, B, C, D, E* and *P* respectively. The points *A, B, C, D* and *E* are joined to *P* by means of straight lines. The line *PE, PD, PC, PB* and *PA* become the radii for the 75°, 60°, 45°, 30°, and 15° parallels respectively. With these radii and *'N'* as centre draw the concentric circles and mark the meridian at interval 15° with the help of a protractor and complete the projection (Fig. 7.14).

(xiv) Gall's Projection

It is a modified cylindrical map projection in which the cylinder cuts

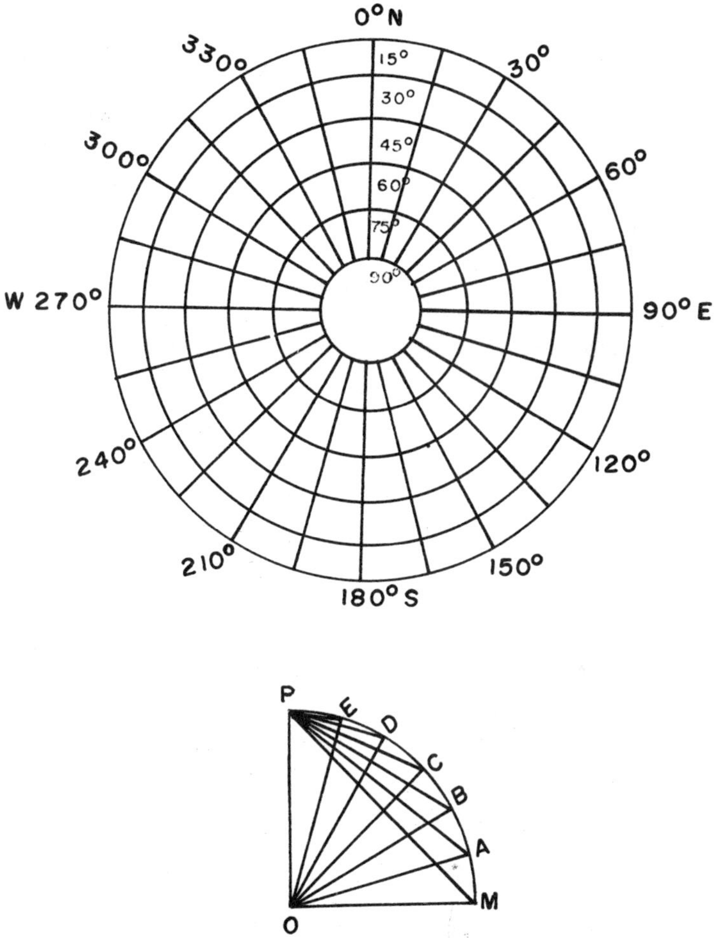

R.F. 1:150,000,000

FIG. 7.14: Zenithal Equal-area projection

the surface of the earth along parallels 45°, N and S. These parallels are made their true to scale length. Meridians are equi-spaced. Parallels and meridians drawn as straight lines and intersect at right angles to each other. However, it is not an equal-area projection. There is less distortion on higher latitudes.

Properties

1. Parallels and meridians are represented by straight lines and intersect at right angles,
2. Meridians are equi-spaced but parallels are not equi-spaced,
3. Only 45° N and S parallels are drawn true to scale, and
4. It is neither orthomorphic nor equal-area.

Uses: Instead of other cylindrical Projection, it is often used for general maps of the world.

Example: Draw a Gall's Projection for the world on a scale 1:200,000,000 at interval of 15°.

The following calculations will be required for mathematical construction.

Length of the equator = $2\pi R \text{ Cos } 45°$
Spacing between the equator and the parallel = 1.7071 R tan f/2
where: 1.7071 is the constant and ϕ is degree.

Radius (R) = 250,000,000/200,000,000
= 1.25"

Length of the equator = $2\pi R \cos 45°$
= $2 \times 22/7 \times R \times 1.25 \times 0.707$
= 38.89/7
= 5.56"

Distance between equator and 15th parallel = 1.7071 R tan 15/2
= $1.7071 \times 1.25 \times \tan 75°$
= 2.13×0.132
= 0.28"

Distance between equator and 30th parallel = 1.7071 R tan 30/2
= $1.7071 \times 1.25 \times \tan 15°$
= 0.57"

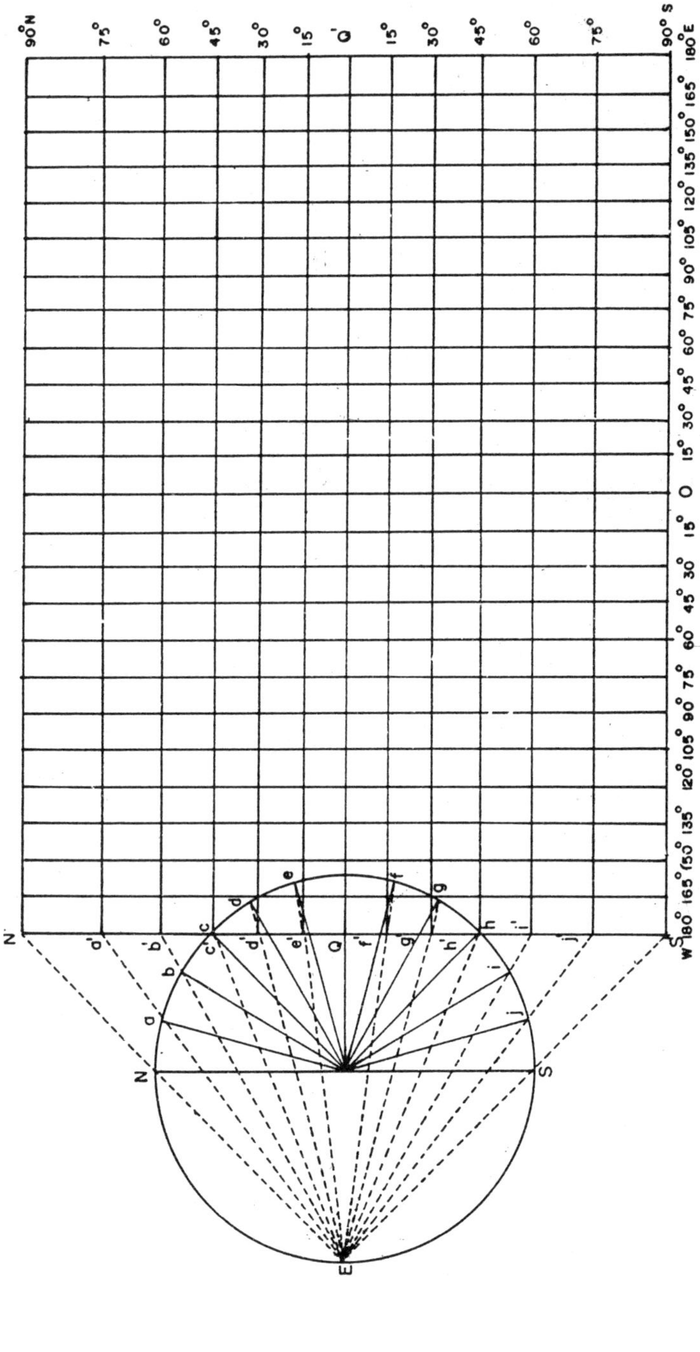

R. F. 1 : 200,000,000

FIG. 7.15: Gall's projection

Distance between equator and 45th parallel = 1.7071 R tan 45/2
= 1.7071 × 1.25 × tan 22.5°
= 2.13 × 0.414
= 0.88"

Distance between equator and 75th parallel =1.7071 × R tan 75/2
= 1.7071 × 1.25 × tan 37.5°
= 2.13 × 0.707
= 1.63"

With these calculations, the Gall's projection can be drawn easily (Fig. 7.15).

Graphical Construction

Draw a circle with the radius *OQ* of 1.25" and extend upto *Q* to represent the equator *AQ'* which is 5.56" in length. Mark *NS* and *EQ* to represent the polar and equatorial diameters respectively. Let the cylinder passes vertically through 45° N and S cutting *EQ* at *A* and *B*. Mark angle 15°, 30°, 45°, 60°, 75° at the centre *(O)* according to the given interval (15°). Let the rays of the light passes from *E* through the various latitudes. Thus *N, a, b, c, d, e, f, g, h, i, j* and *s* are projected to *N', a', b', c', d', e', f', g', h', i', j'* and *s'* along *N'S'* which is equal to the length of meridian. Draw parallel lines to equator to represent parallels from their respective points, which is equal in length to that of 45th parallel (5.56"). Now divide the equator into 24 division and draw perpendiculars to get the meridians. Thus, complete the projection (Fig. 7.15).

7.4 CHOICE OF PROJECTION

When we transfer the various parts of the globe over a plane surface, we are confronted with some difficulties and we have to choose among various projections which serve distinct specific purposes. Some projections truly represent the equatorial lands and others limit itself to the polar areas. Besides that, some projections are true to area, some to shape and direction.

Let us first take a look at the problems that arise while making the choice of map projections.

1. Where the projection is required to project the globe as a whole or a part of it and what is the size and extent of the

area for which the graticule is needed.

2. The purpose of the map.
3. Whether it is easier to calculate and to draw the graticule.

The first problem implies the consideration of the following:

(i) The world maps, continuous or in two hemispheres (i.e. North and south Hemisphere).
(ii) Region extending north and south of the equator, such as Africa, south America, Indian, Pacific and Atlantic oceans.
(iii) Regions with large extent north and south of the equator e.g. China. U.S.A., Australia.
(iv) Regions with their greater extent near the poles.
(v) Small countries.
(vi) Topographical maps.

For the purpose of construction of the world maps, the main projections generally used in classes are:

(1) Zenithal, (2) Mollweide's elliptical, (3) Mercator's, (4) Cylindrical equal area, Gall's, and (5) Sanson-Falmsteed.

Among these projections, Zenithal projection is used for the world in two hemispheres, while others are used for continuous world maps.

The choice of Zenithal for the world map lies in the fact that two maps are needed for showing the hemispheres and the result will be that there will be less distortions in this than in the map in which the world is shown continuously. For showing the wind systems, occean currents, isobars etc. the continuous world maps are useful. Sanson Falmsteed projection may also be rejected because there is much more distortion along the margins of the map than in the other three projections. Besides, for the distribution maps of the world, the choice will be restricted to a few because only homolographic projection will be most suitable for them. If we have to. show the distribution of tropical and sub-tropical crops like rice, cotton, tobacco, sugarcane etc. on the world map, then cylindrical equal area projection will be more useful than Mollweide's and sinusoidal as in case of cylindrical Equal Area projection within 30° of the equator, distortion is much less and the meridians and parallels intersect each other at right angles so the shape is relatively much more correct. In addition to this, it is very easy to construct this projection. On the other hand, if the distribution of temperature, crops like wheat is required to be shown on the world map, Mollweide's and Sinusoidal projections would prove to be better than the cylindrical as in case of cylindrical, there is much distortion than the Sinosoidal. For communication and navigation maps, Mercator's projection is very important.

For an extensive region like Africa which stretches equally north and south of the equator, Sanson Falmsteed may provide the most appropriate graticule, which combines good shape with truly proportional arc, though Zenithal and Cylindrical Equal area are also used for it. For other are like the Indian, Pacific and Atlantic oceans, our choice may rest upon Equatorial Zenithal equidistant or equal area projection or stereographic Normal Zenithal in which the equator is a straight line passing across the country.

North America and Eurasia in northern hermisphere and Australia and Oceanica in southern hemisphere lie, more or less, entirely north and south of the equator respectively. For these we may select between conical with two standard parallels, Bonne's and Polyconic. Zenithal projections are suitable for these regions whose length and breadth are equal because the amount of accuracy decreases equally in all directions as the graticule is drawn from central point along a line which passes vertically through the centre. For polar maps, Zenithal are therefore as ideal as cylindrical for equatorial lands, other things being equal.

A modified form of cylindrical projection is used for topographic survey maps.

Bonne's Projection has been generally applied for India and Asia by Atlas markers because the difficulty of greater dimensions along the base does not arise because of the curved meridians.

8

SURVEYING

8.1 DEFINITION

Surveying is the art of making observations, measurements and representations in order to determine the relative position of objects on the earth's surface and producing what is known as the map. The map which shows only the positions of the various objects relating to each other is known as Planimeter map in which curvature of the earth is not taken into consideration.

There are two types of surveys :

1. Plane survey, and
2. Geodetic Survey.

8.2 CLASSIFICATION

Surveys are classified according to (a) The purpose, (b) The instruments used, and (c) The place where it is done.

According to purpose surveying may be characterised as :

(1) Topographical Surveys
(2) Cadastral Surveys
(3) Geodetic Surveys
(4) Engineering Surveys
(5) Railway Surveys
(6) Military Surveys

1. *Topographical Survey*

These surveys are conducted to include natural features of the country such as rivers, hills, lakes etc. as well as some of the artificial features such as roads, canals, bridges, towns etc. with a view to make the resulting map understandable to an ordinary person.

2. *Cadastral Survey*

These surveys include boundaries of fields, building etc. The scale of this survey is comparatively larger than topographical maps.

3. *Geodetic Survey*

These surveys take into account the curvature of the earth considering it as a sphere or a spheroid. It is conducted with very precise instruments.

4. *Engineering Survey*

These surveys are conducted to obtain the data for any type of projects, such as roads, railways etc. From these data computation and estimates of various projects are prepared.

5. *Railway Survey*

These surveys are conducted to determine the best possible route for locating the railway line.

6. *Military Survey*

These surveys are conducted to determine the best positions of attack and defence from military point of view.

Surveys are also classified according to the instruments used. They are usually divided into four groups.

- (i) Chain Survey (Chain or tape is used)
- (ii) Compass Survey (Prismatic Compass or Sextant is used)
- (iii) Levelling Survey (Dumpy level is used)
- (iv) Theodolite Survey (Theodolite is used)

(i) Chain Surveying

Chain and tape surveying is the simplest mode of survey. It is done with the help of chain and tape, so it is known as chain survey.

There are mainly following types of chain, which are used of linear measurement.

(1) 30 meter chain
(2) 20 meter chain
(3) Steel Band
(4) Gunters chain

1. *30 Meter Chain*

The length of the chain from one extreme end to the other extreme end of the handles is 30 meters. It is divided into 150 equal parts. Each part is known as link and is equal to 20 cms. The links are made of steel wire and are connected by 2 or 3 rings. Distance between the centre of one ring of a link to the centre of the other ring of the link is exactly equal to 20 cms. At the end of every 25 links a brass tag is fixed to denote distance of 5 m, 10 m, 20 m and so on. The central tag is either circular or oval in shape. The tags are marked with letter 'मी' and 'm'. At the extremities of the chain, there are two brass handles, which are attached with several joints. The handle forms part of the end link. The length of the chain is marked over the handle (Fig. 8.1). The 30 meters chain is used for the following purposes.

(i) For engineering surveys such as roads, railways, cities etc. where meter is taken as unit.
(ii) For measuring distance in case of levelling.

2. *20 Meter Chain*

Its construction is exactly the same as that of the 30 meters chain, except that its total length is equal to 20 meters and is divided into 100 links, which is used to determine the equivalent horizontal distances along sloping grounds as in case of hills and embankments by method which is known as *stepping method.*.

Advantages

(i) There is no possibility of the sagging due to its short in length.
(ii) It is easy to carry in hilly regions, because of its light weight.

3. *Steel Bands*

It is more precise than the metallic tapes because it is made of steel, which do not suffer from any permanent strain during working. It vary in length from one meter to hundred meters. It also vary in width from 3 mm to 13 mm. It is either obtained in hard leather cases or wound on some form of metal reel. It is also known as steel tape.

The tape is marked on one side only with a line at every 5 mm, every centimeter, decimeter and meter. The height of marking lines and its colours are as follows:

	Height in mm	Colour/vision
5 mm	3.00	Black line
Every cm	4.00	Black line
Every 5 cm	Full width	Black arrow
Every dm	Full width	Black line
Every meter	Full width	Red line

4. *Gunters Chain*

It was invented by an English Astronomer Mr. Gunter and hence the name. It resembles as engineering chain in every respects except its length, which is equal to 66 feet. It is divided into 100 equal parts. Each part is known as link and is equal to 0.66' = 7.92". The length of 80 gunters chains is equal to one mile and 10 squares chain is equal to one acre.

Tapes

The tapes is used for taking right-angled offsets to the object to be located. The zero end of the tape being held along the objects to be located and the other end being so adjusted as to be perpendicular to the chain line. Care should be taken that the tape is tight while taking the offsets.

The length of tape varies from 20 m to 50 m or 50 ft to 100 ft. It is folded with the help of brass hook which serves as the handle and enclosed in leather cover for safety. There are some common tapes, which have different name due to its different mode of formation (Fig. 8.1b).

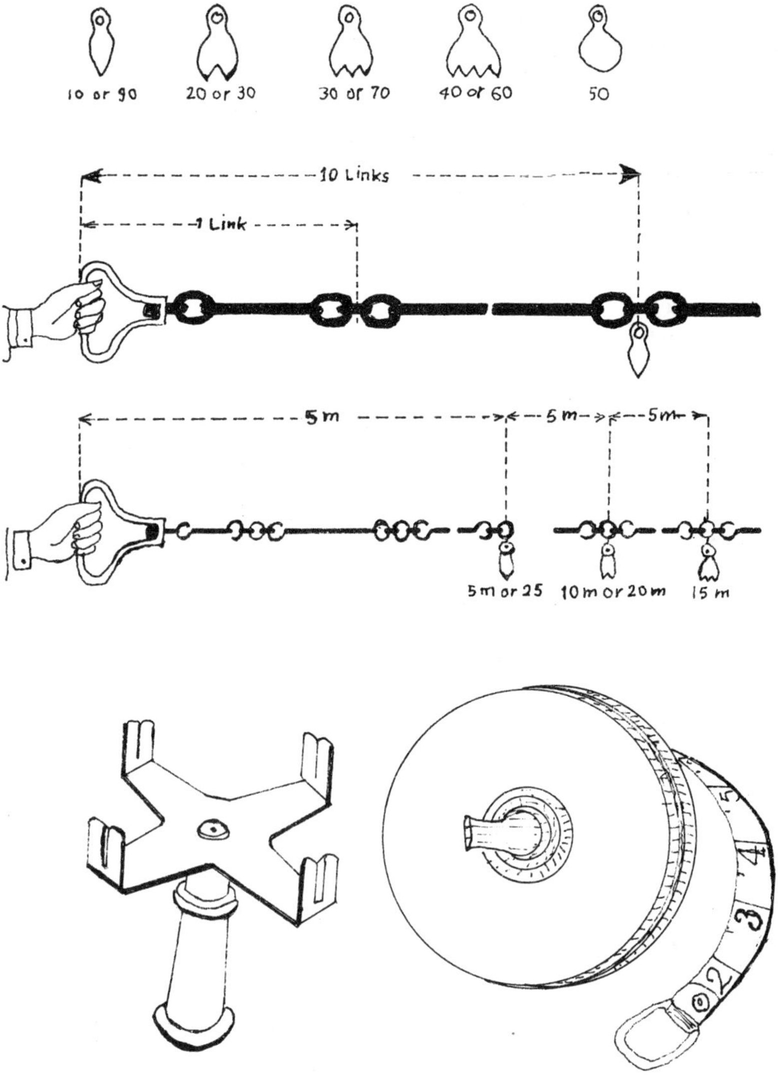

CROSS STAFF

TAPE

FIG. 8.1 Tag, Engineer's Chain, Matric Chain, 8.1(a) Cross Staff, 8.1(b) Tape

(a) *Cloth Tape*

It is made of entirely cloth which has seasonal effects, due to which it suffers from permanent strain during constant pulling on site. This type of tape is likely to get shortened if allowed to dry after being wetted in water. Therefore, for accurate surveying this type of tape is not preferred.

(b) *Metallic Tape*

It is usually made of both cloth or linen and metal wires. In this type of tape small metallic wire 5 to 6 in numbers are inserted in the linen or cloth to prevent the tape from permanent elongation during constant use.

(c) *Steel Tape*

It is made of steel blade and have a very little effect of seasons. There is no any permanent strain during constant pulling on site. Therefore, it is considered to be the best type of tapes. It is generally used to get highest degree of possible accuracy.

Ranging Rod

It consists of either wooden poles or steel rods painted either white, black or red colour with the flag attached to the top of it. Its length varies from 6 feet to 10 feet. The alternate painting is done with a gap of 1 foot to aid the visibility. It is used for ranging lines and also for making the points between which the linear measurement are to be taken. Persons holding the ranging rods are named 'flagmen'. (Fig. 8.2).

FIG. 8.2: Ranging Rod

Arrows

It is made of steel wire measuring about 35 cms long, pointed at one end and coiled into a ring on the other end. Generally 10 arrows are used with chain to record chain length. It is also known as marking pins.

Flags

It is usually made of wooden rods, 2 meters high, painted black and white at every 20 cms. It is used to mark the station points.

Magnetic Compass

It is used for determining the position of magnetic north. It consists of magnetized needle free to rotate in a horizontal plane. One end of the needle points towards the magnetic north pole.

Optical Square

It is made of a thick brass plate fitted with two small mirrors which are at right angles to the plane of the instrument but are inclined at 45 degree to each other. It is generally used for setting out right angles for the measurements of offsets.

Principle

In chaining we measure only horizontal distances and a triangle is the only geometrical figures which can be constructed with the lengths of the three known sides. Thus, the principle of chaining is triangulation, because, it involves the divisions of the field into a framework of triangles.

Procedure

Initially a preliminary checking of the area should be made and the objects which are to be fixed should be marked with wooden pegs. After that, a rough sketch of the area should be drawn to remove the confusion relating to the position of the objects at the time of plotting.

Before starting the survey, the following considerations should be kept in mind.

(i) Chain line should close to the boundary of the area to be surveyed.
(ii) Obstruction like ponds, lakes, hedges etc. Should be avoided for chaining.
(iii) The points selected should be few in numbers so that the triangle formed should be large.

(iv) The triangles formed should be equilateral.

Apparatus Required

For conducting a chain survey on the field, the following apparatus are required.

(i) Chain duly tested.
(ii) 10 arrows.
(iii) 5 to 6 flags or ranging rods.
(iv) Metallic tape.
(v) Wooden Hammer
(vi) Optical square or cross staff.
(vii) 5 to 6 wooden pegs.
(viii) Field book, pencil, knife and eraser.

The field work consists of following operations:

(i) Making a reconnaisance of the area.
(ii) Marking of station.
(iii) Locating the positions.
(iv) Ranging survey lines and fixed objects.

Procedure for Chaining

(i) Fix flags on main station points and wooden pegs on the tie points.
(ii) Stretch the chain straight on the ground along the base line or chain line from the starting point.
(iii) Now take the perpendicular distances from the object to be located to the base line with the tape and the chain link is also noted down at those points.
(iv) Draw the sketch of the objects to which the effects are taken on the correct side of the base line and records the reading.

In this way, all objects are measured from the base line. A field book is prepared with the help of the distance of objects and their chain links. If there are other base lines, follow the same steps, with the help of given scale and field book, draw a final map in laboratory.

For better understanding go through the example of an area as given in (Fig. 8.3).

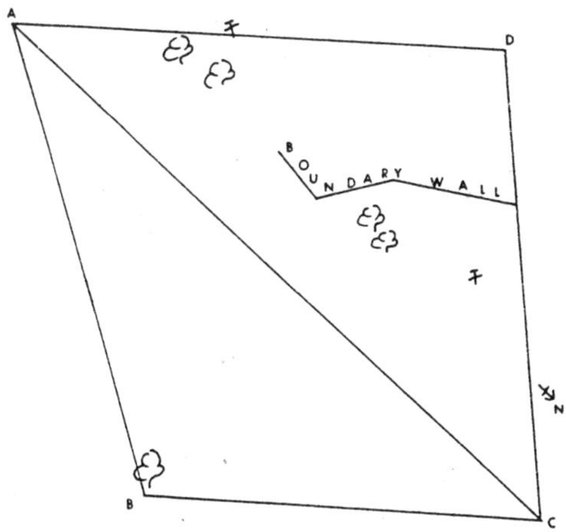

FIG. 8.3: Sketch of the Area for Chain and Tape Survey

Procedure

Divide the area into triangles, Δ ABD and Δ BCD. Assume *AB* a base line. Start the work from point A. Fix a ranging rod at point B. Sketch the chain along the line *AB*. Take the direction of north with the help of trough compass. Give the number of objects systematically for the convenience. Now measure the offset distance at the right angle from the base line as well as note down the chain link as in the following field book.

Field Book

Distance measured in the chain surveying.

Chain Link	Objects	Offset Distance in feet	
A to B			
0	—	—	—
10	1 Left	9'10"	9.8'
15	2 Left	15'8"	15.7'
25	3 Left	12'6"	12.5'
40	4 Left	10'9"	10.8'
50	5 Left	9'7"	9.6'
65	6 Left	16'8"	16.7'
70	7 Left	12'7"	12.6'

(Contd.)

75	8 Left	11'5"	11.4'
75	9 Right	30'4"	30.3'
90	10 Left	25'3"	25.3'
95	11 Left	13'8"	13.7'
100	—	—	—

Such kinds of field books are prepared for *AC, AD, BC,* and *CD,* base line. Now prepare the final map with the help of this field book and the given scale.

The above field book is plotted on the scale 1" = 15'. Firstly prepare a diagonal scale with the help of the given scale. According to this scale, draw a 6.66" line which represents the 100 ft. Mark all offset distances to their respective chain links and complete the map as shown in Fig. 8.4.

(ii) Plane Table Survey

It is the most universal instrument for surveying. It is suitable for topographical work on a small scale. It is a graphical method of surveying in which the field work and its plotting go hand in hand on the field. For the plane table survey the following equipments are generally necessary:

(i) The plane table with a tripod stand.
(ii) Alidade.
(iii) Chain or Tape.
(iv) Spirit level.
(v) Trough Compass.
(vi) Ranging Rods.
(vii) Plumb-bob.
(viii) Drawing essential: paper, pencil, rubber, drawing pins, pencil sharpener, scale etc.

Plane Table

It is a light flat drawing board made of well seasoned wood mounted on tripod. The size varies from, say, 50 cm × 35 cm × 2 cm 65 cm × 50 cm × 2 cm, to 90 cm × 65 cm × 2 cm The board is attached to the tripod by means of central screw with a wing-nut. It may be clamped in any desired position with the help of win-nut. It may be clamped in any desired position with the help of wing-nut (Fig. 8.5).

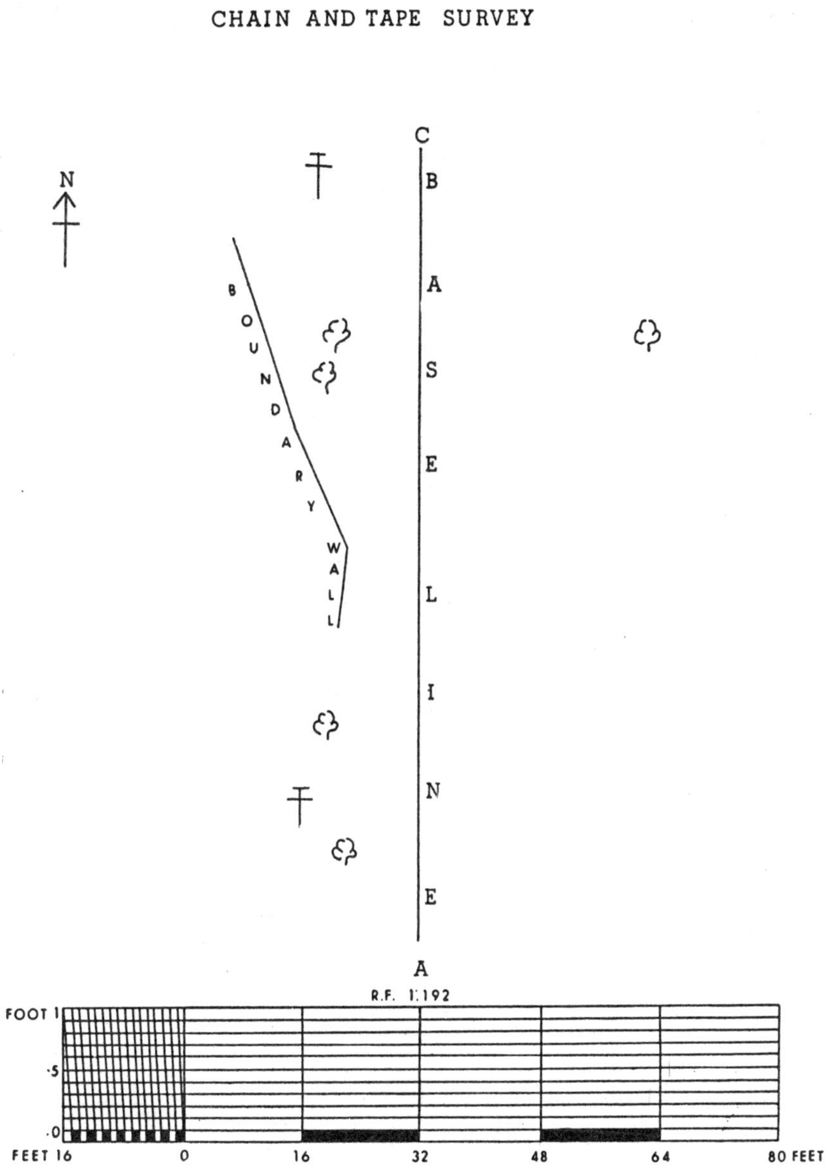

FIG. 8.4: Plotting of field work

FIG. 8.5: Plane Table, Tripod and Alidade

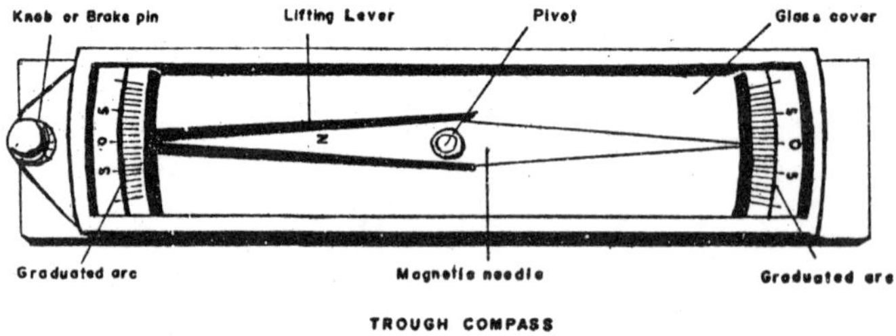

FIG. 8.6(a) : Trough compass

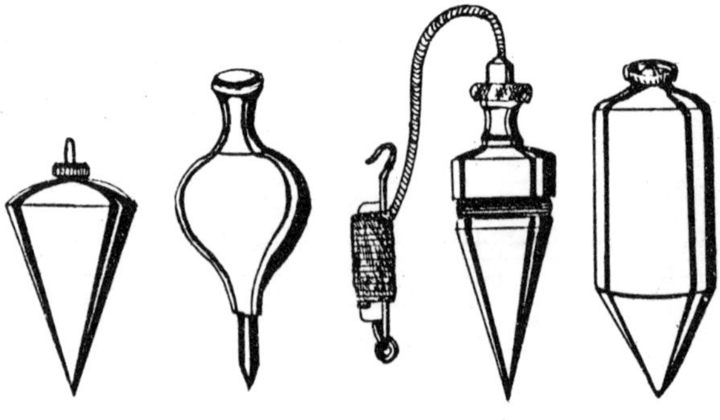

FIG. 8.6(b) : Plumb Bob

Alidade

It is used for sighting and transferring the line of sight to the paper fixed on the board. It consists of a plane ruler about 45 cm long with fiducial edge and is provided at each and with an ordinary sight vane. One of them has a slit (Eye vane) and other has a vertical thread (Object vane) (Fig. 8.5).

Trough Compass

It is used to show the direction of magnetic north. The best form is a long 13 to 15 cms. bar needle pivoted in the centre and enclosed in rectangular box, protected by a glass on its top (Fig. 8.6).

Spirit Level

It is used for levelling the table-top. It consists of a glass tube with spirit and air bubble. When the air bubble rests in the centre, the table is levelled horizontally (Fig. 8.6).

Plumb Bob

It is used to centre the plane-table over the station. It consists of a metallic U-shaped fork, bob and a plumb (Fig. 8.6).

Tripod

It consists of three legs of woods whose height varies from 3 ft to 4 ft. The lower part of the legs are made pointed. The upper part of the legs are joined together to facilitate in fixing the plane table on the top (Fig. 8.5).

There are two procedures, which are to be adopted while working with plane table.

(1) Setting up a plane table over a station point.
(2) Orienting a plane table.

(1) *Setting up a Plane Table*

While setting up a plane table the following points should be observed

(i) Spread the tripod legs and plant them firmly in the ground.
(ii) Take any point on the paper to represent a station point on the ground and fix a pin over it.
(iii) Take a plumbing fork and place its point over the mark on the paper.
(iv) Now shift the table bodily until by plumbing the point on the paper is practically over the station point and level the board with the help of spirit level.
(v) Place the trough compass on the plane table and take the magnetic north by coinciding the needle point to 0-0 of the graduated are provided at the ends of the compass.

The plane table is now said to have been set up over the station point and the plane tabler can proceed with the work and locate objects.

(2) *Orienting a Plane Table*

A plane table is said to be oriented when it is levelled and clamped in such a position that each line plotted on the paper corresponds in direction with the line on the ground it represents. In order to orient a plane table over a station point whose position is already plotted on the paper, proceed as before upto the step (iv) described in setting up a plane table and then further. (v) Place the edge of the trough compass parallel to the magnetic meridian line already drawn on the paper and loosen the wing nut screw. Now rotate the plane table in azimuth till the ends of the magnetic needle of the trough compass point to 0-0 of the graduated scale at ends of the compass. Clamp the wing nut screw and the table is said to be oriented.

Methods of Surveying

Generally, there are following methods of working with plane table :

(i) Radiation.
(ii) Traversing.
(iii) Intersection.
(iv) Resection.

The following are some of the factors to be considered in selecting stations for a plane table survey :

(i) The triangle formed with the adjacent stations to be well conditioned i.e. no angle of a triangle should be less than 30 degree nor more than 120 degree.

(ii) Each station should be clearly visible from the adjacent stations.

(iii) The length of sight to be neither too long nor too small.

(a) *Radiation Method*

It is more suitable when the area is small and plane. In this method the plane table is oriented in any one convenient position from which all points to be located can be seen. After the rays are drawn to the objects to be located by the help of alidade, the corresponding points on the plane table are fixed according to given scale by actually measuring the distances to the objects.

In the Fig. (8.7) let *A, B, C, D,* be four station points on the ground. Set up the plane table at *'O'* from where all points to be located are visible. Clamp the plane table and with the help of trough compass draw the North and south line at one corner of the paper. Direct the alidade towards station A and draw a ray. Now measure the distance *Oa* on the ground and to any suitable scale mark off a length *'Oa'* along the ray on the paper and fix the position of *'A'* on the paper. Similarly locate station *B, C,* and *D,* mark off the length *Ob, Oc,* and *Od* along their respective ray on the paper and fix the position of *B,C* and *D* on paper. In this way all stations and details are located.

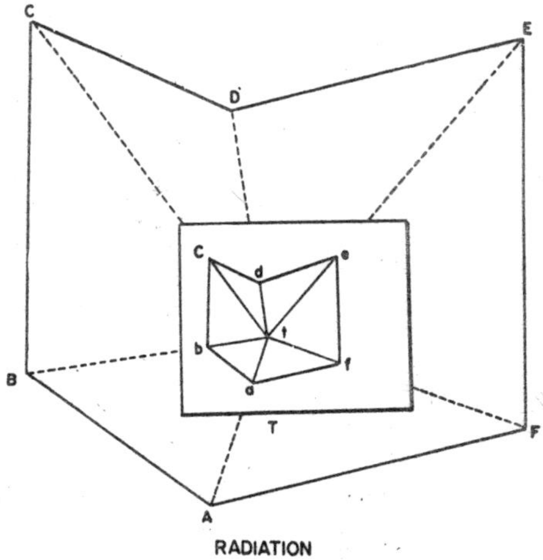

RADIATION

FIG. 8.7: Radiation method

(b) *Traversing Method*

This method consists in shifting the table at each station orienting and measuring the lengths between stations. The details are filled in by the radiation method.

In the Fig. 8.8. Let *A, B, C, D* be four stations selected for the plane table survey. The sequence of operation is as follows :

(i) Set up the plane table over station *'A'* and mark a point 'a' on the plane table, which is exactly over station *'A'* on the ground and clamp the plane table.

(ii) Keeping one edge of the alidade touching point *'a'* direct it to intersect station *B, D* and for check to station *'P'* and draw ray *ab, ad* and *ap.*

(iii) Measure the distance *AB, AD* and *AP* on the ground and to any suitable scale mark off the length *'ab', 'ad'* and ap along the ray on the paper and fix the position of station *B, D,* and *P* on paper.

(iv) Shift the plane table from station *'A'* and set it over station *'B'* in such a way that the point *'b'* on paper is exactly above station *'B'* on the ground. Place the alidade along the line *'ba'* and rotate the plane table till looking through the alidade station *'A'* is intersected. Clamp the plane table.

(v) With alidade touching *'b'* direct it to intersect station *'c'*. If the plane table is correctly oriented at station *'B'*, then looking through the alidade and intersecting station *'p'* a ray through *'b'* to *'p'* will pass through *'p'* on the plane table.

(vi) Measure *BC* on the ground, mark off *'bc'* on the paper and locate the position of *'c'*.

(vii) Shift the plane table from station *B* and set it up over station *'C'* so that *'c'* on paper is vertically above station *'C'*. Place the alidade along line *'cb'* and revolve the plane table till station *'B'* is intersected. Clamp the plane table.

Now with the alidade touching stations *'C'* direct it to *'D'* and proceed as before.

(c) *Intersection Method*

In this method objects or stations are located by drawing two rays from extremities of a measured base line plotted on the paper. The intersection of these two rays to the same objects locate the required point. This method approximate and is used when the objects are inaccessible.

In Fig. 8.9 let *AB*, be two station point, draw a base line *'ab'* on the paper to any convenient scale to represent line *'AB'* on the ground.

(i) Set up the plane table over station *'A'* and centre it so that the point *'a'* is vertically over station *'A'* on the ground.

(ii) Orient the plane table by laying the alidade along *'ab'* and turn the table until line of sight cuts point *'B'*. Clamp the plane table and draw N-S line on one corner of the paper with the help of trough compass.

(iii) With the alidade touching *'a'* sight the various objects required to be located. Suppose locate the corners of a building *M* and *N*. Draw rays to *M* and *N*.

(iv) Shift the plane table and set it over station *'B'*. Centre the table by keeping *'b'* vertically above the station *'B'* and orient it by laying the alidade along 'ba', rotate the plane table until the line of sight cuts point *'A'*. Clamp plane table.

(v) Draw rays through *'b'* to the same corners *M* and *N*. The intersection of these two with the respective two previous rays locates the positions of two corners *'m'* and *'n'*.
In this way any detail can be located.

(d) *Re-section Method*

If there are two or three well defined points on the field whose relative points are already plotted on the paper, the problem consists in locating on the paper the position occupied by the plane table on the field by observing the two or three visible points. If observations are made to two visible points, the problem is called two-point-problem. Whereas if observations are made to 3 known points the problem is called three-point-problem. (Fig. 8.10)

There are following four methods of solving this problem:

(i) Mechanical tracing paper method.
(ii) Triangle of errors method.
(iii) Graphical or Bassels method.
(iv) Geometrical method.

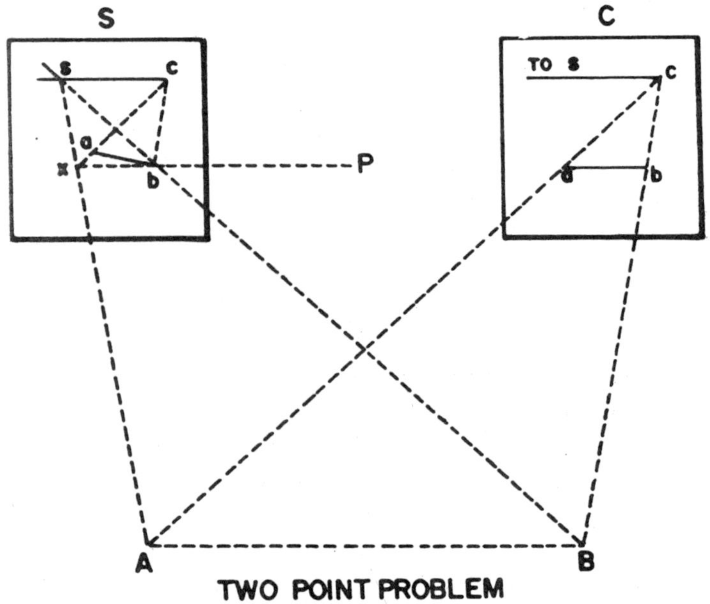

FIG. 8.8: Traversing Method

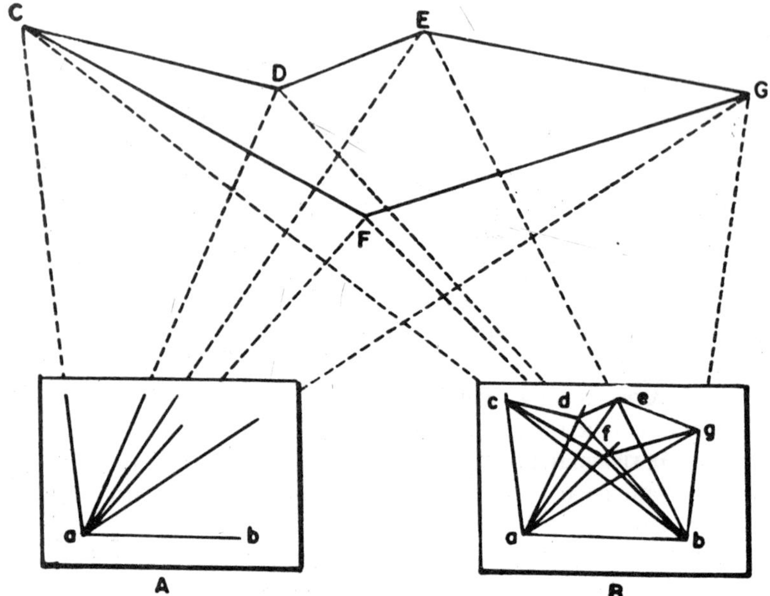

FIG. 8.9: Intersection Method

Tracing Paper Method

In this method a tracing paper of the size of plane table is taken and the following procedure is adopted.

(i) Orient the plane table with the eye or by trough compass and clamp it.
(ii) Suppose *a, b* and *c* be the corresponding plotteu position of the station *A, B* and *C* on the plane table (Fig. 8.10).
(iii) Place the tracing paper over the plane table.
(iv) Take any point '*X*' on the tracing paper and draw rays from '*X*' to objects *A, B* and *C* respectively.
(v) Now move the tracing paper over the paper on the plane table until the three rays *XA, XB* and *XC,* pass through points *a, b* and *c* respectively, when this is done mark with a pin the point *X* to leave impression on the paper below. That point is the required position '*Y*' of the plane tables occupying on the field.

Set the alidade along *Ya* and then turned the table until *Ya* is directed to '*A*'. If the plane table is properly oriented, the rays *Xb* and *Xc* will intersect station *B* and *C* respectively along rays *Xb* and *Xc* respectively.

If due to little error the three rays do not meet at *X* a triangle of error will be formed which will usually be so small that a slight movement of the board will correctly orient the p'ane table.

After confirming the table position, take Y_2 as base line and repeat the whole processes of intersection to include more objects on the sheet.

(iii) Prismatic Compass Survey

Prismatic compass is the main instrument for this type of survey. It is invented by Captain Karter in 1814. The main parts of the prismatic compass are shown in Fig. 8.11.

Prismatic compass is an angular instrument used for measuring magnetic bearings of a line. It is the most convenient and portable form of instrument, generally filled on a tripod. It is provided with a small right angled prism. There is an eye slit at one end of the compass and an object vane provided with a horse hair attached to its centre for intersecting the objects at the other end of the compass. There is a dial

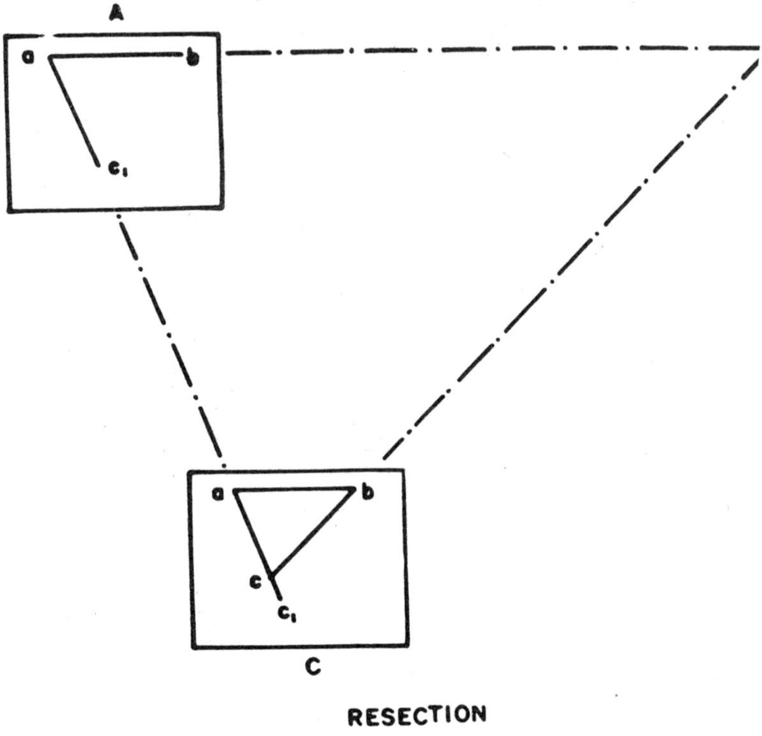

FIG. 8.10: Re-section Method (Tracing Paper Method)

graduated into 360 degrees, which is supported on hardened steel pivot. The graduation of the dial are written in inverted and they get erected after reflection through the prism. The least reading is 30 minutes and by interpolation by eye it can be read to 15 minutes and even upto 7.5 minutes. A small magnetic needle which shows north and south direction is attached below the dial.

The greatest advantage of prismatic compass is that both sighting the object as well as reading circle can be done simultaneously without changing the position of the eye. The circle is read at the reading at which the hair line appears to cut the graduated ring.

The plotting of the objects on the map are done with the help of angles (measured through the compass) and distance between them. There are two methods of prismatic compass survey.

(a) *Open Traverse*

In this method, surveyor does not come back to the starting station. It is also known as route surveying. In this method, the bearings of the lines are calculated on the basis of the bearings of that line which has a difference of 180 degrees in its fore and back bearings. If, however, there is no such line in which the two bearings differ by 180 degrees, the correction should be made from the mean value of the bearing of that line in which there is least discrepancy between the back sight and fore sight readings.

Example: The following bearings were observed while transversing with a compass. Determine the corrected bearings and plot the traverse to 1 cm = 20 meters.

Line	Length in meter	Forward bearing	Backward bearing
AB	100	40°50'	220°50'
BC	50	92°25'	272°10'
CD	60	35°45'	215°20'
DE	40	110°25'	290°35'

Solution

It is observed that difference between back and fore bearings of the line *AB* is exactly 180 degree. Hence station *A* and *B* and free from any attraction and all other bearings measured at these stations are also correct. Thus, the observed bearing of *BC* (i.e. 92 degree 25') is correct. The correct bearing of *CB* will, therefore, be 92 degree 25' + 180 = 272 degree 25' while the observed bearing is 272 degree 10'. The error at C is therefore – 15' and a correction of + 15' must be applied to the bearings measured at *C*. The correct bearing of *CD* thus become 35 degree 45' + 15' = 36 degree and that of *CD* as 36 degree + 180 degree = 216 degree. The observed bearing of *DC* is 215 degree 20'. Hence, the error at *D* is 215°20' – 216° = –40' and a correction of +40' must be applied to the bearings measured at *D*. The correct bearing of *DE* thus becomes 110 degree 25' + 40' = 111 degree 5' and that of *ED* as 111 degree 5' + 180° = 291 degree 5' (See Fig. 8.12).

The result may be tabulated as under:

Line	Length in Meter	Observed bearing	Correction	Corrected bearing
AB	100	40°50	0 at A	40°50'
BA		220°50	0 at B	220°50'
BC	50	92°25	0 at B	92°25'
CB		272°10'	+15' at C	272°25'
CD	60	35°45'	+15' at C	36°00'
DC		215°20'	+40' at D	216°00'
DE	40	110°25'	+40' at D	111°05'
ED		290°35'	+30' at E	291°05'

(b) *Close Traverse*

In this method, surveyor come back to the starting station. In other words, the last station coincides with the starting station, thus completing a closed polygon. In this method, the bearing of the lines are calculated in the same way as the bearing of the open traverse are calculated.

Example : The following are the bearings taken on a closed compass traverse. Determine the corrected bearings and plot the traverse to 1 cm = 5 m, scale.

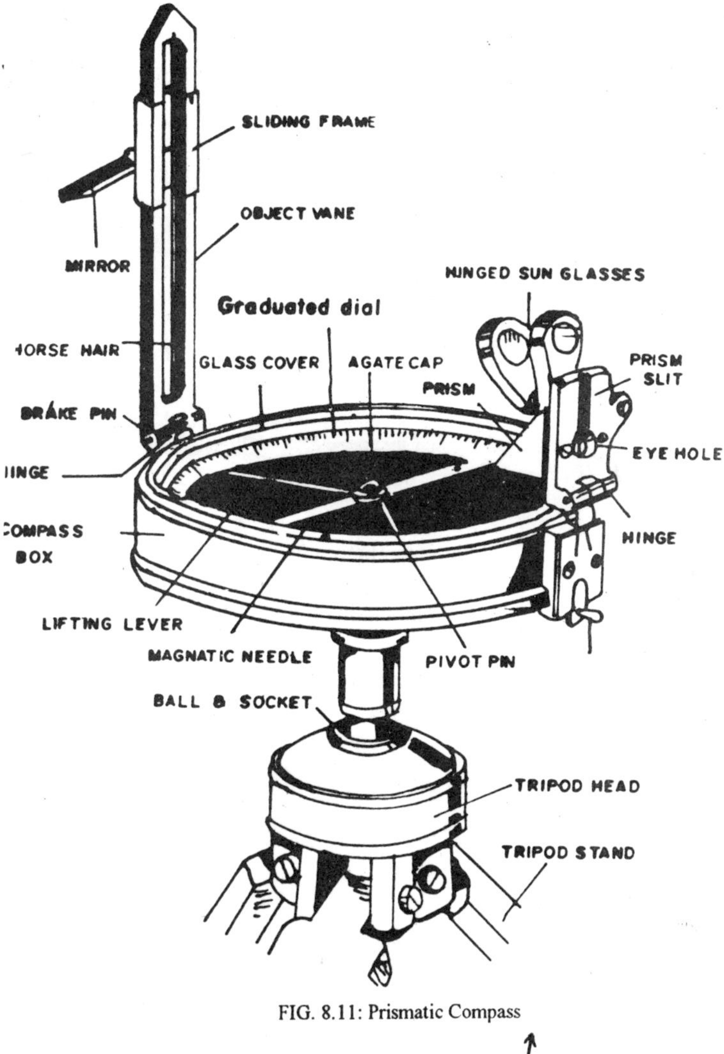

FIG. 8.11: Prismatic Compass

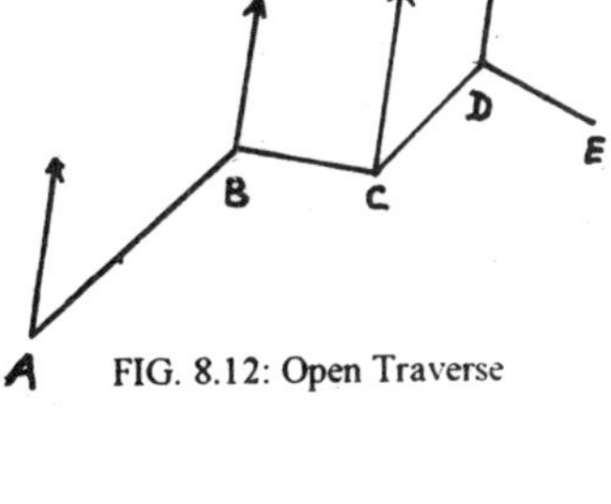

FIG. 8.12: Open Traverse

Line	Distance (in meter)	Forward bearing (FB)	Backward bearing (BB)
AB	27	51°15'	231°15'
BC	32	120°30'	300°30'
CD	33	212°35'	32°35'
DE	25	269°40'	89°45'
EA	25	342°50'	162°45

Solution

The result may be tabulated as under, following the procedure of solution of open traverse.

Line	Distance (in meter)	Observed bearings	Correction	Corrected bearings
AB	27	51°15'	O at A	51°15'
BA		231°15'	O at B	231°15'
BC	32	120°30'	O at B	120°30'
CB		300°30'	O at C	300°30'
CD	33	212°35'	O at C	212°35'
DC		32°35'	O at D	32°35'
DE	25	269°40'	O at D	269°40'
ED		89°45'	–5 at E	89°40'
EA	25	342°50'	–5 at E	342°45'
AE		162°45	O at A	162°45'

Plotting

Take a point as station *A*. Draw a vertical line representing north south at station *A*. Mark an angle of 51° degree 15' (Corrected bearing of *A*) in the clockwise directions. Mark station *B* by taking a distance of 2.7 cm (representing 27m). Likewise mark all stations with the help of their corrected bearings and distance between them according to scale (1 cm = 10m).

Thus, in Fig. 8.13, polygon *A,B',C',D',E'* represents an unbalanced traverse having an error equal to *AA'*, that is known as "closing error". Distribute that total closing error *AA'* linearly to all the sides in proportion to their lengths by a graphical construction shown in Fig. 8.14. In Fig. 8.14 *AB',B'C', C'D'* and *E'A'* represent the length of the sides of the traverse to the reduced scale (it can be represented either to the same scale or to the reduced scale). At *'A'* draw a vertical line *aA'* equal to the length *AA'*. Join *'A'* with *'a'*. At *B'*, *C'*, *D'* and *E'*, draw vertical lines and extend upto *Aa* line, the corresponding errors *B'b*, *C'c*, *D'd* are found. Mark the distance *B'b*, *C'c* and *D'd*. E'e parallel to *AA'* on *B'*, *C'*,*D'* and *E'* station and finally join them by a broken line to complete the plotting. The polygon *ABCDE* so obtained (shown in broken line) represent the adjusted traverse.

Errors in Compass Survey

The errors may be classified as:

(a) Instrumental errors
(b) Personal errors
(c) Errors due to natural causes.

(a) *Instrumental Errors*

They are those which arise due to the faulty adjustments of the instruments. They may be due to the following reasons.

(i) The needle not being perfectly straight.
(ii) Pivot being bent.
(iii) Sluggish needle.
(iv) Blunt pivot point.
(v) Improper balancing weight.
(vi) Plane of sight not being vertical.
(vii) Line of sight not passing through the centre of the ring.

(b) *Personal Errors*

They may be due to the following reasons.

(i) Inaccurate levelling of the compass box.
(ii) Inaccurate centering.
(iii) Carelessness in reading and recording.
(iv) Inaccurate bisection of signals.

(c) *Natural Errors*

They may be due to the following reasons.

(i) Variation in declination.
(ii) Local attraction due to proximity of local attraction forces.
(iii) Magnetic changes in the atmosphere due to clouds and storms.
(iv) Irregular variations due to magnetic storm etc.

(iv) Levelling

Levelling is an operation for determining the relative elevation of various points on the surface of the earth. It directly determines the attitudinal differences between neighbouring points. The elevation of any point referred to some plane of reference which may be mean-sea-level or any assumed datum is known as absolute elevation of a point.

(a) *Patterns of Levelling Instruments*

Level, that are used for this work, are of several types. The old and traditional levels are the Y-level and the Dumpy level. Out of which Dumpy level is commonly used (Fig. 8.14). Among the new types, Watt's High Way level is accurate and also simple to operate.

It is essential to know some of the definitions used in levelling work.

Bench Mark

This is a permanent point which may be the plinth of a building or any permanent structure whose elevation with respect to some datum is already known. The bench marks are always used for starting the levelling work.

Ordinance Bench Mark

In this case the plane of reference is situated at Liverpool and is a plane 1.42435 m above the old Dock still at Liverpool.

Datum

It is an imaginary level surface to which all the elevations of various points are referred to.

Staff Point

It is the point on the surface of the earth where the levelling staff is placed, whose elevation is required to be determined.

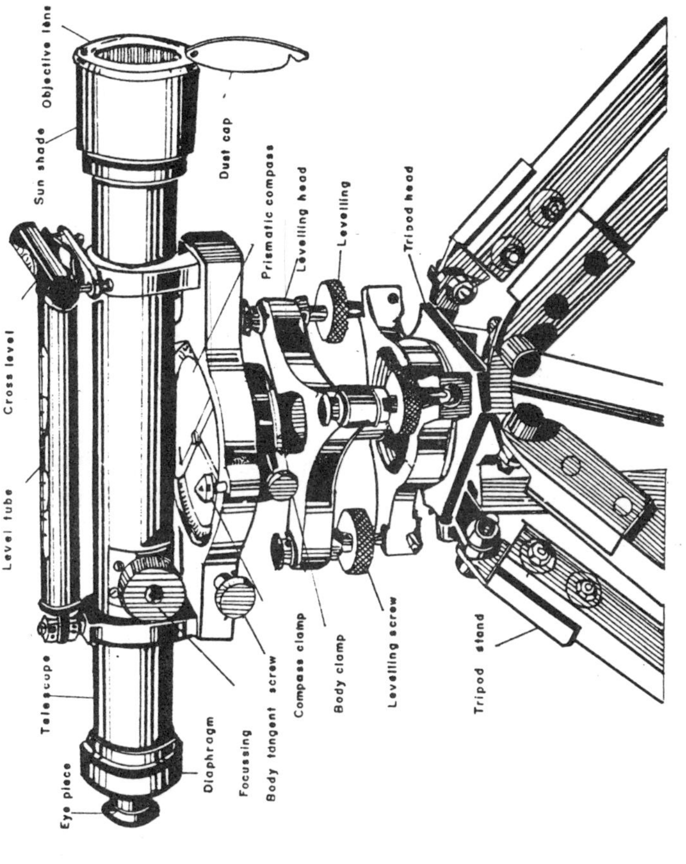

FIG. 8.14: Dumpy Level

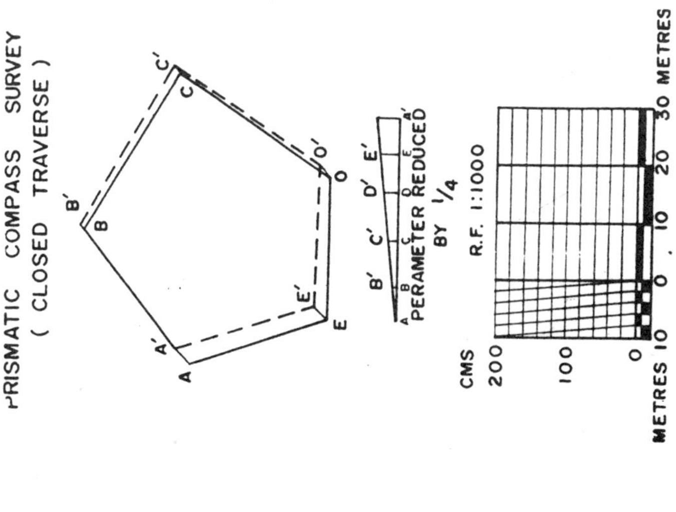

FIG. 8.13: Close Traverse

Staff

The staff is made of well-seasoned teakwood 75 mm wide and 18 mm thick. Its length is about 4 meters folding type or 3 meters single. The zero reading on the staff is at the bottom. Each meter is divided into 10 parts, each part is equal to 0.1 meter i.e. 1 decimeter. One decimeter part is divided in 20 divisions. Each divisions therefore is equal to 0.005 meter i.e. 5 mm. The divisions are alternately pointed in black and white. All the full meter divisions are marked *'M'* and pointed in red through the total width of the staff. At the end of each decimeter are marked 1, 2, 3, etc. in black which actually denotes 0.1, 0.2, 0.3 etc. of a meter. For example the reading on the staff at position marked will be read as 0.55 (Fig. 8.15a). Hence the least count of the smallest divisions that can be read on the staff is 0.005 meter or 5 mm.

The feet staff's length is about 20 ft. On the rectangular cross-section distance are marked in feet, tenth of feet and hundredths of feet. Hence the least count is 0.01 feet (Fig. 8.15b).

(b) *Objects of Levelling*

The following are the objects for which levelling is done:

(i) To determine the profile of a country for selecting the best site for the location of road, railway or canal.
(ii) To determine, in case of drainage projects, the positions of the bridge and valley lines and calculating the drainage area for designing suitable sizes of sewers.
(iii) To determine the undulations of the site proposed to be reserved for constructing a building.
(iv) To establish some supplementary bench marks at far off distances.
(v) To set and to check the formation levels of roads and canals.

(c) *Differential Levelling*

When levelling is carried on to find out the substantially differential heights of various points, it is known as differential levelling. The procedure is to fix instrument in between two points and find out the fore-sight and back-sight. After that shift the instrument in between another two points and find out the foresight and backsight. In this way complete the whole levelling.

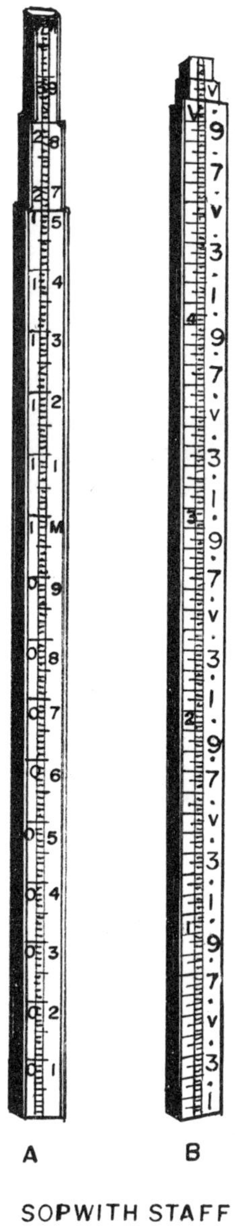

FIG. 8.15: Staff

(d) *Profile Levelling*

By the help of this method a number of readings are taken from one position of the instrument to determine the undulating nature of the terrain along a predetermined line. In this method the rise or fall in elevation at any point is determined with respect to this preceding point.

Let *ABCDE* be several points on the profile of a ground. Level is set up at the point *'X'*. After levelling the instrument, staff readings are taken on point *A, B, C, D,* and *E*. Now, suppose the staff reading on the point *'A'* is 3 m and that on the point *'B'* is 2 m it shows that the point *'B'* is 1 m higher than the point *'A'*. In other words there is a rise of 1 m at the point *B* above its preceding point *'A'*. Now suppose the staff reading kept at the point *'C'* reads 4 m. This shows that at the point *'D'* there is a fall of 4 – 2 = 2 m with respect to its preceding point *'B'*. In this way take remaining readings, calculate rise and fall, and their reduced level as shown in the following table:

Station	Dista- -nce in meters	Back Sight	Interm- -ediate Sight	Fore Sight	Rise	Fall	Reduced Level (R.L.)
A	0	3.000	—	—	—	—	100.000
B	100	—	2.000	—	1.000	—	101.000
C	150	—	4.000	—	—	2.000	99.000
D	200	—	3.000	—	1.000	—	100.000
E	250	—	—	2.000	1.000	—	101.000
Airthmetic Check		3.000	—	2.000	3.000	2.000	Last R.L. –First R.L.

Diff.= 3.000 – 2.000 = 1.00 Diff.= 3.000 – 2.000 = 1.00 Diff. = 101.00 – 100.00 = 1.00

(e) *Plotting of the Profile*

When heights of various points along a continuous line are known, they can be plotted to give a profile or section. In plotting the profile two scales are used: a horizontal scale and a vertical scale (Fig. 8.16).

8.3 THE SEXTANT

The sextant is based on the principle that when a ray of light is

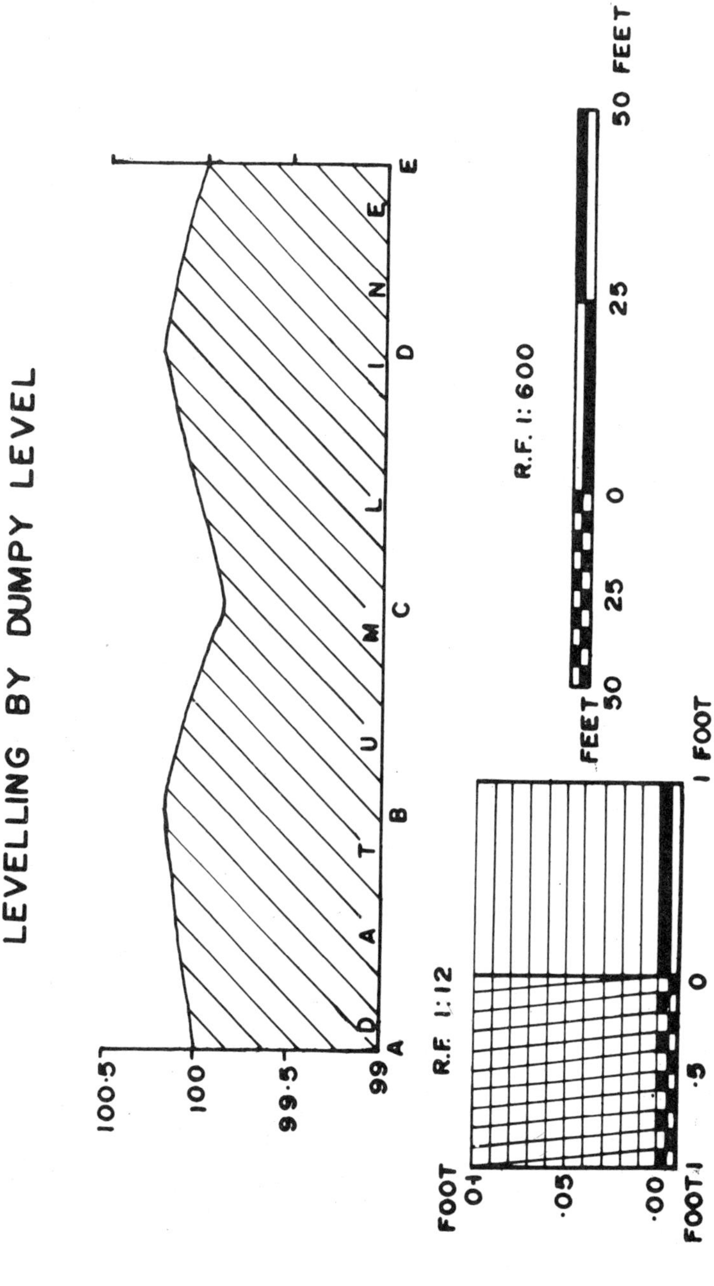

FIG. 8.16: Plotting of the Profile

reflected successively from two mirrors, the angle between the first and the last directions of ray is twice the angle between the planes of the two mirrors.

There are mainly three types of Sextant.

(1) Box Sextant.
(2) Nautical Sextant.
(3) Sounding Sextant.

(v) Box Sextant

The box sextant is small pocket instrument used for measuring horizontal and vertical angles, measuring chain angles and locating inaccessible points (Fig. 8.17).

A box sextant consists of following parts.

(i) A circular box about 8 cms in diameter and 4 cms high.
(ii) A fixed horizon glass, silvered at lower half and plain at upper half.
(iii) A movable index glass fully silvered.
(iv) An index arm pivoted at the index glass and carrying a vernier at the other end.
(v) An adjustable magnifying glass to read the angle.
(vi) A milled-headed screw to rotate the index glass and the index arm.
(vii) An eye hole or a telescope for long distance sighting.
(viii) A pair of coloured glass for use in bright sun.
(ix) A slot in the side of the box for the object to be sighted.

Measurement of Horizontal Angle

1. Hold the instrument in the right hand and bring the plane of the graduated arc into the plane of the eye and the two points to be observed.
2. Look through the eye hole at the left hand object through the lower unsilvered portion of the horizon glass.
3. Turn the milled-headed screw slowly so that the image of the right hand object after double reflection, is coincident with

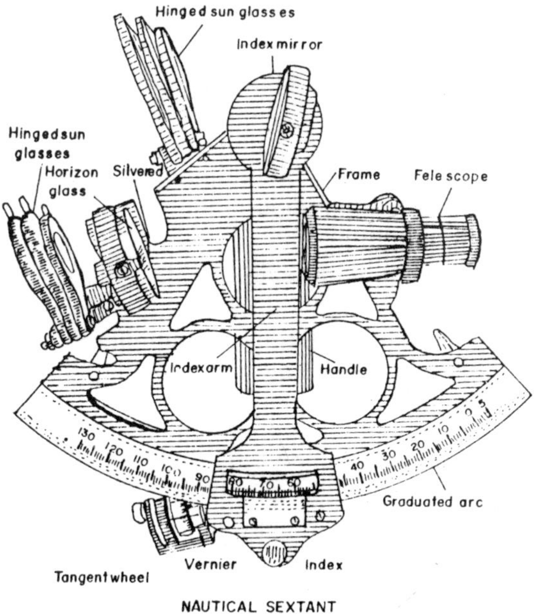

FIG. 8.17: Box Sextant

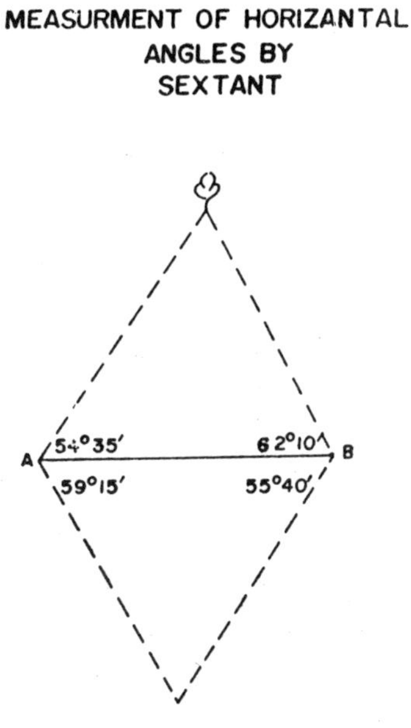

FIG. 8.18: Plotting of the field work

the left-hand object, view directly through the upper half of the horizon glass. Clamp the vernier. If a slow motion screw is provided, bring the images of objects into exact coincidence.

The reading on the vernier given directly the angle (Fig. 8.18).

Measurement of Vertical Angle

Vertical angles may be measured by holding the sextant so that its arc lies in a vertical plane. If it is required to measure the vertical angle between two points, view the lower object directly and turn the milled headed screw until the image of the higher object appears coincidents with the lower one.

(vi) Indian Clinometer

It is developed by Survey of India, so it is known as Indian clinometer. It is also known as tangent clinometer. It is used to find out the heights of various objects as well as for contouring. The parts of an Indian clinometer are shown in Fig. 8.19.

Working

1. Place the instrument on a plane surface (Plane table) which is planted over a station point and is approximately horizontal.
2. Bring the bubble of the level tube in mid-run by means of adjusting screw.
3. Keep the flaps perpendicular and turn up the steets which keep the flaps in position.
4. Look through the eye hole, slide the pinion up or down till the far distant object is intersected by the horizontal crosswire of the sliding window and clamp it in that position.
5. Read the angle of inclination on the left hand side of the graduated vertical vane.
6. Read the tangent value corresponding to the angle of inclination on the right hand-side of the graduated vertical vane.
7. Measure the horizontal distance of the object from the eye hole of the instrument.
8. Now measure the height of the eye hole above the peg.
9. Finally calculate the height of object and add the height of eye hole to it to get the actual height of the object above peg.

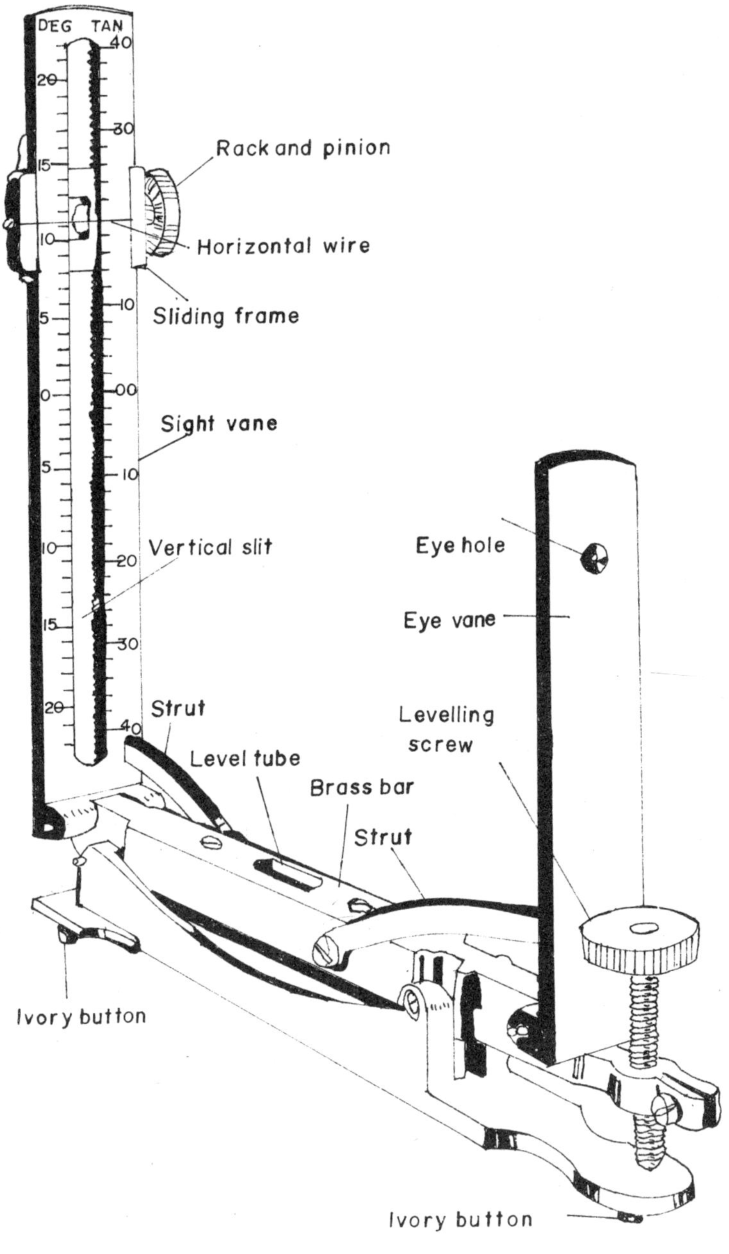

FIG. 8.19: Indian Clinometer

The above illustration deals with the case when the object is above peg. When the object to be sighted is lower, the procedure is the same, only the difference is that the height of the instrument above the peg should then be subtracted. For further details go through examples (Fig. 8.20).

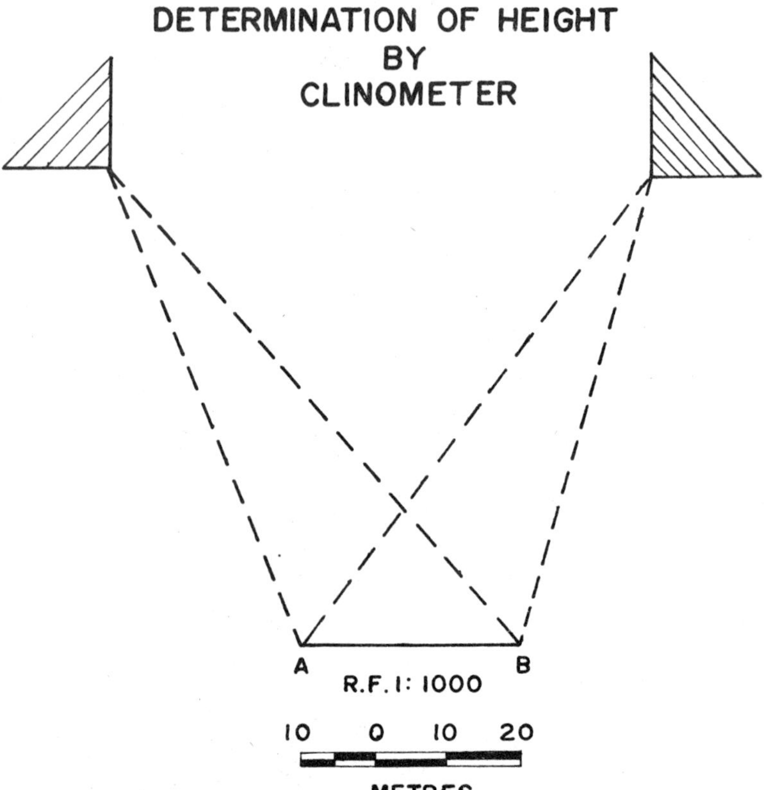

FIG. 8.20: Plotting of Field Work

(vii) Abney's Level

An Abney's level is a small handy level. It is one of the forms of handy clinometer which can be very usefully employed for contouring a small area as well as measuring angles in vertical plane. It consists of an open tube made of brass, square in section, entirely without lenses, whose line of sight is marked by the pin hole eye piece at one end of the tube and a horizontal wire at the other end. There is a small transparent spirit level over the top of the tube, rigidly fastened to a vernier arm which moves over a scale of degree, graduated on the side of the semicircular arc with zero in the centre. There is a polished silvered surface inclined at 45 degrees to the line of sight to reflect the bubble to the eye. It occupies half the section of the tube, leaving the other half for sighting object. On turning the thumb screw, the bubble of the spirit level is intersected by the cross-wire, while looking through the eye hole. Vernier can read correct to 10 minutes (Fig. 8.21)

Determination of Relative Heights or Slope by Abney Level

1. Hold the Abney level in the right hand and sight through the eye hole and intersect the object by the horizontal wire fixed inside the half portion of the object end.
2. With the left hand turn the thumb screw until the refection of the bubble is cut centrally by the mark on the mirror.
3. Then lower the instrument carefully and read the vernier which represents the angle of inclination of the tube with the horizontal. Suppose the angle is equal to 15 degree 30'.
4. Now determine the distance of object from eye hole as well as the height of the eye hole from ground. Suppose the distance is about 100 meters and height is about 1.6 meters.
5. Take the tangent value of degrees (15 degrees 30') from the tangent table, which is about 0.2772.
6. Put the tangent value, horizontal distance and height of eye hole in the following formula.

Horizontal distance x tan ϕ + Height of the eye hole.

$$= 100 \times 0.27732 + 1.6$$
$$= 27.732 + 1.6$$
$$= 29.332$$
$$= 29.33 \text{ meters}$$

Thus, the height of object from station is 29.33 meters. When the object to be sighted is lower, the height of the eye-hole should then be subtracted.

7. The supposed angle is 15 degree 30' (15.50') the levelled edge will read slope of 45:15.5 (1:0.34).

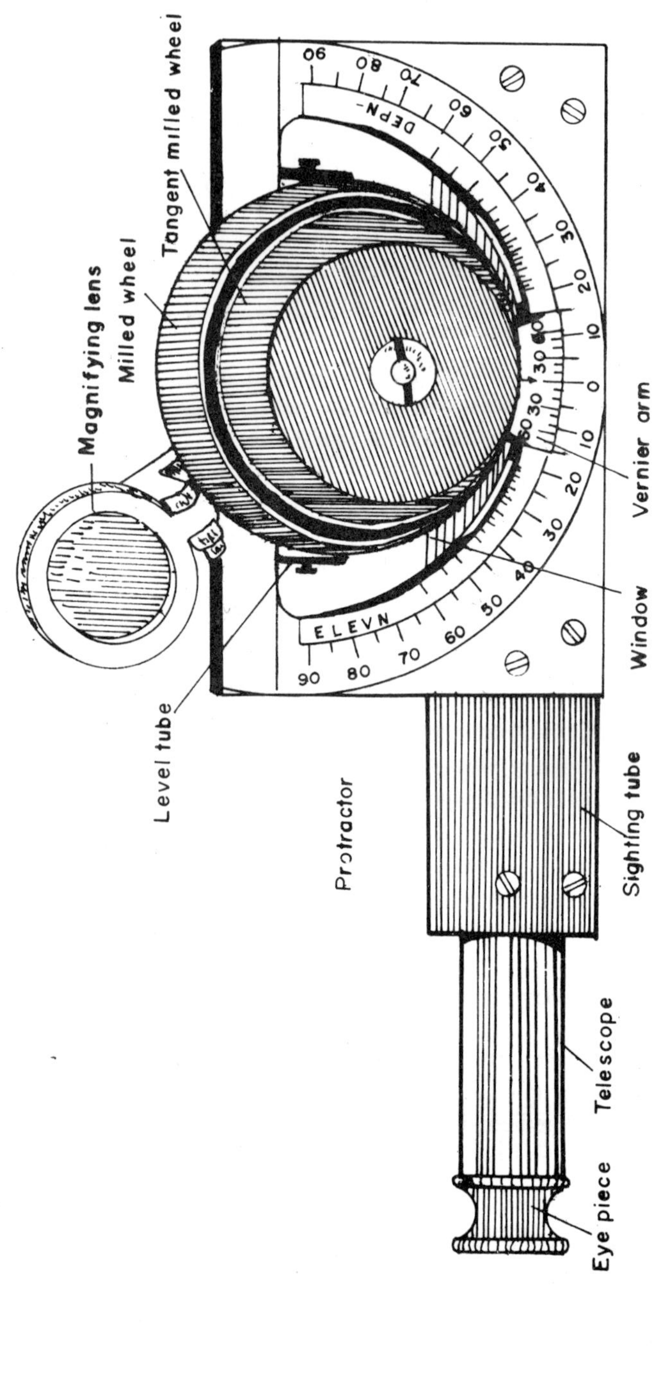

FIG. 8.21: Abney's Level

(viii) Theodolite

This instrument is used to measure the vertical and horizantal angles with great precision, specially in case of Triangulation Survey, Traverse Survey. On the basis of precision, there are various types of theodolite. Some of the important are as follows:

(a) Transit Vernier Theodolite
(b) Double Arc Theodolite
(c) Watt's Micrometer Theodolite
(d) Tavistock Theodolite
(e) Wild T-2 and T-3 Theodolite

These theodolites differ in respect of least count i.e. the least count of T-2 and T-3 Theodolite is one second, whereas, the least count of Transit vernier theodolite is 20 seconds. The principle in all types of theodolite is the same.

Transit Vernier Theodolite

Transiting means capable of being rotated along the transvers axis. In this theodolite, the telescope is capable of revolving rpund its transverse and hence the name transit theodolite (Fig. 8.22). The size of instrument depends on the diameter of the main scale graduated plate. The scale is graduated to read degrees. Each degree is divided into three equal parts, so that each part denotes 20 minutes. The minimum reading (20 minutes) on main scale is subdivided into 60 equal parts, so that one part denotes 1/60th of 20 minutes or 20 seconds on vernier scale. Thus the least count of the instrument is 20 seconds.

Procedure

There are three observations at the time of field survey by theodolite.

(i) field sketching and selction of stations.
(ii) Setting up the instrument.
(iii) Measurement of angles.

(i) Field Skectching

Firstly, prepare a rought sketch of the area which is to be surveyed because this help in the preparation of final maps.

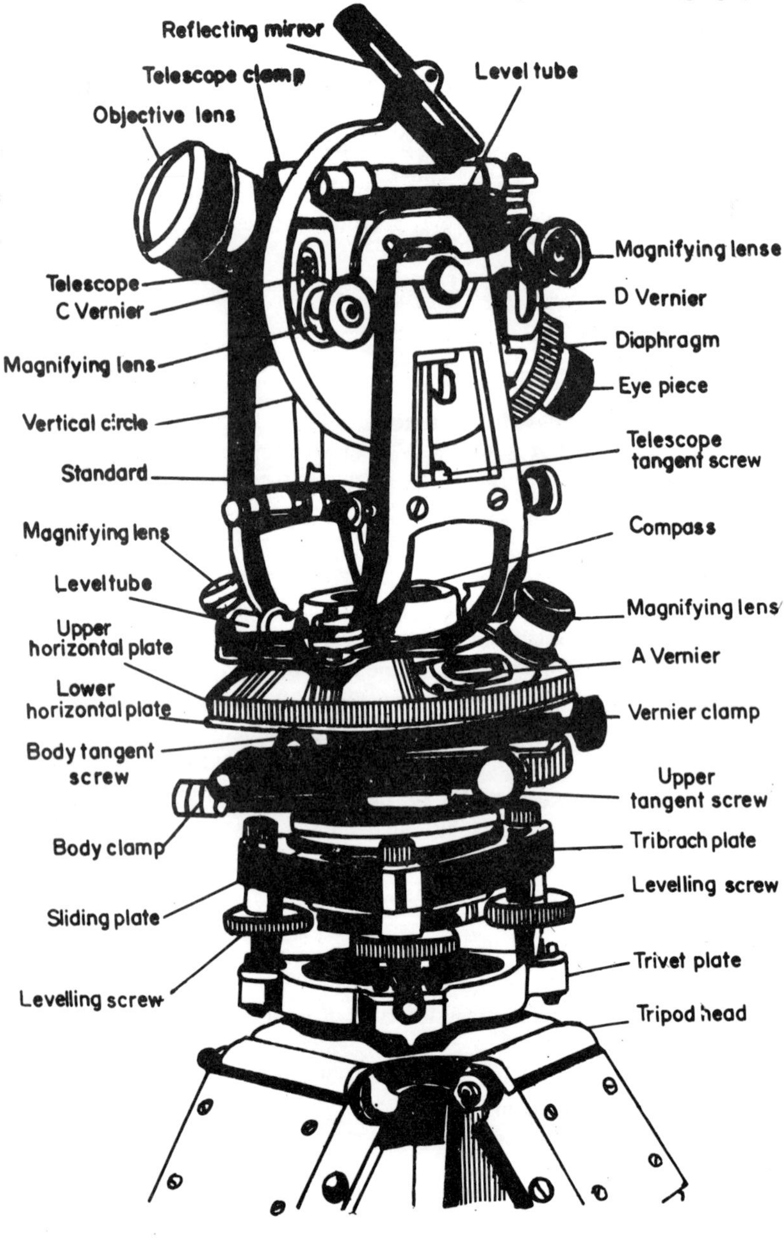

FIG. 8.22: Theodolite

(ii) Setting up the instrument

Firstly, centering the instrument with the help of legs of tripode and levelling screw by centering the bubble of the sprit level by giving the two screw inward or outward motion simultenuously. Then put the telescope perpendicular to third screw and by moving the screw bring the bubble in centre. This process is repeated till the bubble remain in the centre of the sprit level.

(iii) Measurement of Angles

Firstly, make the zero setting by coinciding the zero of Vernier *A* to 360° of main scale and zero of vernier *B* to 180° of main scale. The start the procedure to take the angular distance between the objects.

For measuring the horizantal angle, firstly left hand object is sighted and object is seen at the cross hair of the diaphram. Now the lower plate is claped and vernier plate released. Now, systematically measure the angular distance one by one by correctly sighting the object with the help of the vernier tangents. The readings at two verniers are noted and the mean gives the required angle.

9

STATISTICAL METHOD

9.1 MEASURES OF CENTRAL TENDENCY

Measures of Central Tendency generally comprises of following types:

(i) Mathematical Average

(a) Arithmetic Average or Mean.
(b) Geometric Mean.
(c) Harmonic Mean.

(ii) Average of Position

(a) Median.
(b) Mode.

(i) Arithmetic Average or Mean

The arithmetic mean is an average value of a given series lying in between the highest and lowest values of the series. It is obtained by adding the values of all observations of series and dividing the total by the number of observations. This method can be expressed in the form of mathematical formula.

Suppose the values of observations of a series are respectively X_1, X_2, X_3,...X_n and their arithmetic average is represented by $\bar{X}$ then,

$$\bar{X} = \frac{X_1 + X_2 + X_3 + \ldots\ldots X_n}{N}$$

$$= \frac{\Sigma X}{N}$$

Where;

$\bar{X}$ = Arithmetic Average
X = Value of observations
Σ = Total or summation
N = Number of items.

Characteristic

1. The value of the arithmetic mean is determined by every item in the series.
2. It is greatly affected by extreme values.
3. The sum of the deviations about it is zero.
4. The sum of the squares of deviation from the arithmetic mean is less than of those computed from any other point.

Advantages

1. It is rigidly defined.
2. It is based on all values given in the distribution.
3. It is least affected by fluctuation of sample.
4. It is easily calculated.
5. It is most easily understood.
6. It is most amenable to algebraic treatments.

Disadvantages

1. It may be greatly affected by extreme items and its usefulness as a summary of the whole may be considerably reduced.
2. When the distribution has open-end classes, its computation would be based on assumption, and therefore, may not be valid.

Examples : Calculate the arithmetic average or mean of the following data.

Size of items		
20	40	70
28	50	75
30	55	80
35	60	85

Solution

$$X = \frac{20+28+30+35+40+50+55+60+70+75+80+85}{12}$$

$$= 628/12$$

$$= 52.33 \text{ Ans.}$$

Short-cut Method

The short cut Method is based on the property that the algebraic sum of the deviations of individual observations from the arthmetic mean equal to zero. Thus, the arthmetic mean of 3, 5, 6, 7, and 9 is equal to 6. If the difference of each of these observations from the mean is calculated, it would be –3, –1, +1, +3, their total become zero.

One can assume any arbitrary mean and take the deviation from this assumed mean. If the sum of the deviation become zero then that arbitrary value is arithmetic mean If however, the sum of deviation is not zero, then the actual arithmetic mean is calculated from the following formula.

$$\bar{X} = A + \frac{\Sigma d}{N}$$

where $\bar{X}$ = Actual arithmetic mean
A = Assumed mean
Σd = The sum of deviations from the assumed mean.
N = Number of observations.

Example : Compute the arithmetic mean by short-cut method of 4,5,8,10 and 12.

Solution

Suppose assumed mean (A) = 9

X	d = (X–A)
4	– 5
5	– 4
9	0
10	+1
12	+3
	$\Sigma d = -5$

$$\bar{X} = A + \frac{\Sigma d}{N}$$

$$= 9 + -5/5$$

$$= 9 - 1$$

$$= 8 \text{ Ans.}$$

Calculation of arithmetic mean from a grouped data or discrete series

In a grouped data or discrete series, the value of the mid point of class interval or the value of the variables are multiplied by their respective frequencies. The products obtained in this way are added together. This total value of observations is divided by the number of observations, or items that is equal to total number of frequencies.

Mathematically expressed as

$$= \frac{\Sigma fX}{N} \text{ or } \frac{\Sigma fX}{\Sigma X}$$

Example : Calculate the arithmetic mean from the following data.

Weekly wages (in rupees)	No. of labourers
10 - 20	5
20 - 30	10
30 - 40	8
40 - 50	12

Solution

Weekly wages (in Rupees)	Mid-point	f	fx
10-20	15	5	75
20-30	25	10	250
30-40	35	8	280
40-50	45	12	540
		$\Sigma f=35$	$\Sigma fx=1145$

$\bar{X}$ = Σfx/N or Σf×/Σf
= 1145/35
= 229/7
= 32.71 Ans.

Example : Calculate the arithmetic mean from the following data.

Frequency of Marks of 20 students.

Marks (X)	Frequency (f)
5	2
5	8
10	5
15	4
20	1

Solution

X	f	fx
5	2	10
5	8	40
10	5	50
15	4	60
20	1	20
	Σf=20	Σfx=180

$\bar{X}$ = Σfx/Σf = 180/20 = 9 Ans.

Short-cut Method

Arithmetic mean is also calculated by short-cut method from grouped data or discrete series. In this method the deviations of the value of the mid-point of class interval or the value of the variables from assumed mean are first found out and after that they are multiplied by their respective frequencies. The summation of these products is divided by the total number of frequencies and added to assumed mean. In this way, get the arithmetic mean.

Algebraically : $\bar{X} = A + \frac{\Sigma fd}{N}$

Example : Calculate the Arithmetic mean from the following data.

Weekly wages (in rupees)	No. of Labourers (f)
10 - 20	5
20 - 30	10
30 - 40	8
40 - 50	12

Solution

Weekly wages (in rupees)	Mid value	No. of labourers	Deviations from assumed mean (35) (d)	fd
10-20	15	5	–20	–100
20-30	25	10	–10	–100
30-40	35	8	0	00
40-50	45	12	15	+180
				Σfd = –20

Therefore,

$$\bar{X} = A + \frac{\Sigma fd}{N} = 35 + \frac{-20}{35}$$

$$= \frac{1225 - 20}{35} = \frac{1205}{35} = 34.42 \text{ Ans.}$$

Example : Calculate the Arithmetic mean from the following data :

Frequency of Marks of 20 students

Marks	Frequency
5	2
5	8
10	5
15	4
20	1

Solution

Marks	Deviation from assumed mean (10) (d)	Frequency (f)	fd
5	–5	2	–10
5	–5	8	–40
10	0	5	00
15	+5	4	20
20	+10	1	10
		Σf = 20	Σfd = -20

Therefore,

$$\bar{X} = A + \frac{\Sigma fd}{N}$$

$$= 10 + \frac{-20}{20}$$

$$= 10 - 1$$

$$= 9 \text{ Ans.}$$

(ii) Median

Median is the value of middle item of a series, which must be arranged in ascending or descending order of magnitude. In other words, median divide the series into two equal parts, one part containing values less than the median and the other part containing values more than the median. Median does not take into account the values of all items in a series. Therefore, it is also called positional average.

For ungrouped data median is calculated by using the formula $\frac{(n+1)}{2}$, where, N is the total number of observations.

If however, there are even number of observations in a series then there is no central item dividing the series into two equal parts. For example, if there are 8 observations in a series the median value would be between 4th and 5th, value of 4th observation plus value of 5th observation divided by two.

Characteristics

1. It is an average of position.
2. It is affected by the number of items than by extreme values.
3. The sum of the deviation about the median, signs ignored, will be less than the sum of the deviations taken from any other point.

Advantages

1. It is easily calculated and is not much disturbed by extreme values.
2. It is more typical of the series.
3. The median may be located even when the data are incomplete.

Disadvantages

1. The median is not so well suited to algebraic treatment as arithmetic, geometric and harmonic means.
2. It is not so generally familiar as the arithmetic mean.

Example : Find out the median of the following data.

5, 7, 8, 11, 12, 6, 4, 3, 5.

Solution

These observations would first be arranged in ascending order of magnitude.

Now the series become as follows :

Serial Number	Size of observations
1	3
2	4
3	5
4	5
5	6
6	7
7	8
8	11
9	12

Thus,

Median is in the size of (N + 1)/2
= (9 + 1)/2 = 10/2
= 5th observation

That is :

Median = 6 Ans.

In the above example, the number of observations are odd and there is no difficulty in locating the middle observation, that is 6.

Example : Find out the value of median from the following data

8, 10, 15, 18, 20, 6, 5, 4.

Solution

These observations would be arranged in ascending order as follows :

Serial Number	Values
1	4
2	5
3	6
4	8
5	10
6	15
7	18
8	20

Median is in the size of (N+1)/2
= (8+1)/2
= 9/2
= 4.5th observation

The size of 4.5 observation would be situated in between the observations 4 and 5.

Thus,

$$\text{Median} = \frac{8+10}{2}$$
= 18/2
= 9 Ans.

For grouped observation, median is calculated by using the formula,

$$\text{Median} = L + \frac{(N/2 - C)}{f} \times h$$

Where,

L = Lower limit of class in which median lies.
h = class interval
f = frequency of the class having the median.
C = Cummulative frequency of the next lower class
N = Total number of observations.

To find out the middle observation in this case, firstly find out the value of N/2. The value of N/2 check in the cummulative frequency. After observing the N/2 value, the class interval and its frequency is taken into considerations (see example).

Example : Find out the median value of the following distributions.

Class interval	Frequency
0 - 5	7
5 - 10	12
10 - 15	18
15 - 20	25
20 - 25	15
25 - 30	12

Solution

Class interval	Frequency (f)	Cummulative Frequency (c)
1 - 5	7	7
5 - 10	12	19
10 - 15	18	37
15 - 20	25	62
20 - 25	15	77
25 - 30	12	89
	$\Sigma f = 89$	

The value of N/2 = 89/2 = 44.5, which lies in 15 - 20 class interval. Thus :

$$= L + \frac{\frac{N}{2} - C}{f} \times h$$

$$= 15 + \frac{44.5 - 37}{25} \times 5$$

$$= 15 + \frac{7.5}{5} \times 5$$

$$= 15 + \frac{7.5}{5}$$

$$= 15 + 1.5$$

$$= 16.5 \text{ Ans.}$$

(iii) Quartiles, Deciles and Percentiles

Quartiles are those observations which divide the entire observations into quarters. Quartiles marking the limits of lower and upper quarters are known as 1st and 3rd quartiles respectively. Middle one is called the second quartile which coincides with the median, 5th decile (D_5) and 50th percentile (P_{50}) itself. The quartiles are denoted by Q_1, Q_2, and Q_3.

The quartile observations are ascertained by the following

formulae: for ungrouped data

1st Quartile (Q_1) = (N+1)/4th item

2nd Quartile (Q_2) = $\frac{(N+1)}{2}$ item = Median

3rd Quartile (Q_3) = $\frac{3\,(N+1)}{4}$ item.

The deciles are the nine points which so, divide the series that each portion contains equal number of observations. In this case, the series is divided into 10 equal parts. Symbolically :

1st decile = D_1
2nd decile = D_2
3rd decile = D_3
4th decile = D_4
5th decile = D_5 = Median
6th decile = D_6
7th decile = D_7
8th decile = D_8
9th decile = D_9

The percentile are the ninety-nine points which divide the series into 100 equal parts in such a manner that the parts contain equal number of observations, symbolically :

1st. percentile = P_1
2nd " = P_2
3rd " = P_3
upto 50 th " = P_{50} = Median
upto 99 th " = P_{99}

Example : Find out the values of Q_1, Q_2, Q_3, D_5 and P_{50} from the following given observation, which gives the marks obtained by 20 students in class test.

Roll No.	Marks obtained	Roll No.	Marks obtained
1	33	11	30
2	35	12	45
3	34	13	55
4	25	14	65
5	50	15	70
6	55	16	65
7	70	17	54
8	68	18	52
9	45	19	44
10	35	20	40

Solution

The obtained marks are arranged in ascending order of magnitude

Serial No.	Marks	Serial No.	Marks
1	25	11	50
2	30	12	52
3	33	13	54
4	34	14	55
5	35	15	55
6	35	16	65
7	40	17	65
8	44	18	68
9	45	19	70
10	45	20	70

The value of Q_1, $= \frac{N+1}{4}$

$$= \frac{(20 + 1)}{4}$$

$$= \frac{21}{4}$$

= 5.25th observations

That is, size of 5th observation + 0.25 (size of 6th observation-size of 5th observation)

or 35 + 0.25 (35 – 35)
or 35 + 0.25 × 0
or 35 Ans.

$$\text{The value of } Q_2 = \frac{(N+1)}{2}$$

$$= \frac{(20+1)}{4}$$

$$= \frac{21}{2}$$

= 10.5th observation

That is : size of 10th observation + 0.5 (size of 11th observation - size of 10th observation)

or 45 + 0.5 (50-45)
or 45 + 0.5 × 5
or 45 + 2.5
or 47.5 Ans.

$$\text{The value of } Q_3 = \frac{3\,(N+1)}{4}$$

$$= \frac{3\,(20+1)}{4}$$

$$= \frac{3 \times 21}{4}$$

$$= \frac{63}{4}$$

= 15.75th observation

That is : size of 15th observation +0.75 (size of 16th observation-size of 15th observation)

or 55 + 0.75 (65 – 55)
or 55 + 0.75 × 10
or 55 + 7.5
or 62.5 Ans.

$$\text{The value of } D_5 = \frac{5\,(N+1)}{10}$$

$$= \frac{5(20+1)}{10}$$

$$= \frac{5 \times 21}{10}$$

$$= \frac{105}{10}$$

=10.5th observation

That is : size of 10th observation +0.5 (size of 11th observation— size of 10th observation)

or 45+.5 (50 – 45)
or 45+.5 × 5
or 45+ 2.5
or 47.5 Ans.

$$\text{The value of } P_{50} = \frac{50(N+1)}{100}$$

$$= \frac{50\,(20+1)}{100}$$

$$\frac{50 \times 21}{100}$$

$$= \frac{1050}{100}$$

=10.5th observation

That is : size of 10th observation +0.5 (size of 11th observation— size of 10th observation)

or 45 + 0.5 (50-45)
or 45 + 0.5 × 5
or 45 +2.5 = 47.5 Ans.

Computation of Quartiles, Deciles and Percentiles in a grouped or continuous frequency distribution.

In a continuous series, as in the case of median the values of quartiles, deciles and percentiles lie in the various class intervals and the actual values have to be calculated by the use of algebraic formula, that is:

$$Q_1 = L + \frac{\frac{N}{4} - C}{f} \times h$$

$$Q_2 = L + \frac{\frac{N}{2} - C}{f} \times h$$

$$Q_3 = L + \frac{\frac{3N}{4} - C}{f} \times h$$

$$D_i = L + \frac{\frac{Ni}{10} - C}{f} \times h$$

$$P_i = L + \frac{\frac{Ni}{100} - C}{f} \times h$$

Example : Calculate 1st, 2nd, 3rd quartile, 8th decile and 40th percentile from the following data:

Monthly wages (Rs.)	No. of workers
20 - 25	6
25 - 30	10
30 - 35	15
35 - 40	20
40 - 45	25
45 - 50	30
50 - 55	25
55 - 60	15

Solution

Monthly wages (Rs.)	No. of Workers (f)	Cummulative frequency (c)
20 - 25	6	6
25 - 30	10	16
30 - 35	15	31
35 - 40	20	51
40 - 45	25	76
45 - 50	30	106
50 - 55	25	131
55 - 60	15	146

$\Sigma f = 146$

For Q_1:

The value of N/4 = 146/4 = 36.5
Which lies in 35 - 40 class interval

$$\text{Thus } Q_1 = L + \frac{\frac{N}{4} - C}{f} \times h$$

$$= 35 + \frac{(36.5 - 31)}{20} \times 5$$

$= 35 + 5.5/4$
$= 35 + 1.375$
$= 36.375$
$= 36.38$ Ans.

For Q_2:

The values of N/2 = 146/2 = 73
Which lies in 40 - 45 class interval

$$\text{Thus } Q_2 = L + \frac{\frac{N}{2} - C}{f} \times h$$

$$= 40 + \frac{(73 - 51)}{25} \times 5$$

$= 40 + 22/5$
$= 40 + 4.4$
$= 44.4$ Ans.

For Q_3:

The value of 3N/4 = 3×146/4 = 438/4 = 109.5
Which lies in 50-55 class interval

$$\text{Thus } Q_3 = L + \frac{\frac{3N}{4} - C}{f} \times h$$

$$= 50 + \frac{(109.5 - 106)}{25} \times 5$$

$= 50 + 3.5/5$
$= 50 + 0.7$
$= 50.7$ Ans.

For D_8:

The value of 8N/10 = 8 × 146/10 = 1168/10 = 116.8

Which lies in 50 — 55 class interval

$$\text{Thus } D_8 = L + \frac{\frac{8N}{10} - C}{f} \times h$$

$$= 50 + \frac{(116.8 - 106)}{25} \times 5$$

$= 50 + 10.8/5$
$= 50 + 2.16$
$= 52.16$ Ans.

For P_{40} :

The value of 40N/100 = 40 × 146/100 = 5840/100 = 58.40
Which lies in 40 - 45 class interval

$$\text{Thus } P_{40} = L + \frac{\frac{40N}{100} - C}{f} \times h$$

$$= 40 + \frac{(58.40 - 51)}{25} \times 5$$

$= 40 + 7.4/5$
$= 40 + 1.48$
$= 41.48$ Ans.

(iv) Mode

Mode is a value around which there is the greatest concentration of values. It is the most common value. Generally it is the value which occurs the largest number of times in a series. In the words of Zizek "the value occurring most frequently in a series (or group) of items and around which the other items are distributed most densely".

Mode cannot be calculated in a series of individual observations unless it is converted into a discrete or continuous series.

There are four important methods to estimate mode of a series, which is as follows:

(i) Locating the most frequently repeated value in the series.

(ii) Estimating the mode by calculation.
(iii) Locating the mode by graphic method.
(iv) Estimating the mode from the mean and the median.

Out of the above mention four methods, here (ii) and (iv) are taken into consideration.

Estimating the mode by calculation is made by the use of the following formula

$$\text{Mode} = M_o = L + \frac{f_m - f_1}{2f_m - f_1 - f_2} \times h$$

Where :

L = Lower limit of model class
f_m = Maximum frequency
f_1 = frequency of class preceding to the model class
f_2 = frequency of class succeeding to the model class
h = class interval.

Characteristics

1. It is an average of position.
2. It is not affected by extreme values.
3. It is the most typical value of the distribution.

Advantages

1. Since it is the most typical value it is the most descriptive average.
2. It is easy to locate the approximate mode, but the determination of true mode requires extensive calculations.
3. Since the mode is usually an 'actual value', it indicates the prices value of an important part of the series.

Disadvantages

1. Unless the number of items is fairly large and the distribution reveals a distinct central tendency, the mode has no significance.
2. It is not capable of mathematical treatment.

3. In a small number of items the mode may not exist.

Example : Calculate the value of mode from the following data

Class interval	Frequency
0 — 10	8
10 — 20	6
20 — 30	9
30 — 40	12
40 — 50	7
50 — 60	5
60 — 70	4

Solution

Maximum frequency is 12
Which lies in 30 – 40 class interval
Thus:

$$M_0 = L + \frac{f_m - f_1}{2f_m - f_1 - f_2} \times h$$

$$= 30 + \frac{12 - 9}{2 \times 12 - 9 - 7} \times 10$$

$$= 30 + \frac{3}{24 - 16} \times 10$$

$= 30 + 30/8$
$= 30 + 3.75$
$= 33.75$ Ans.

Estimating the mode from the mean and the median, have the following empirical relationship which is given by Karl Pearson.

Mode = Mean – 3 (Mean – Median)
= Mean – 3 Mean + 3 Median
= 3 Median – 2 Mean

Thus, Mo = 3 median – 2 mean.

Example : If the mean and median of a moderately asymmetrical series are 24.5 and 25.3 respectively, what would be its most probable mode?

Solution

Using the emperical relationship :
Mo = 3 median – 2 mean
$= 3 \times 25.3 - 2 \times 24.5$
$= 75.9 - 49$
= 26.9 Ans.

Just, as the value of mode can be found out in an asymmetrical distribution, similarly the value of mean or median can be found out by the same emperical relationship i.e.

Mode = 3 median – 2 mean

9.2 MEASURES OF DISPERSIONS

(i) Range

It is the simplest measures of dispersions. The range of any series is the difference between the highest and the lowest observation in the series.

Symbolically

Range (R) = L – S

Where :
L = Highest observation
S = Lowest observation.

If the range is divided by some of the extreme observations, the resulting value is known as, *The Ratio of Range* or *The Co-efficient of the Scaller*.

Symbolically

$$\text{The ratio of Range} = \frac{L - S}{L + S}$$

Example : Calculate the range and its co-efficient from the following data.

Year	1987	1988	1989	1990	1991	1992
Loss in '000 Rs.	130	200	250	135	140	300

Solution

$$\text{Range} = L - S = 300 - 130 = 170$$

$$\text{Co-efficient of Range} = \frac{L - S}{L + S}$$

$$= (300 - 130) / (300 + 130)$$

$$= 170/430$$

$$= 0.395 \text{ Ans.}$$

When the available data are in the grouped form or in the form of a continuous frequency distributions, then the range is calculated by using any one of the following method.

(i) By subtracting the lower limit of lowest class interval from the highest limit of the highest class interval.

(ii) By subtracting the mid-value of the lowest class interval from mid-value of the highest class interval.

Example : Calculate the range and its co-efficient from the following data.

Class interval	10 - 20	20 - 30	30 - 40	40 - 50	50 - 60	60 - 70
Frequency	30	15	12	8	6	20

Solution

Class interval	Mid-value	Frequency
10 - 20	15	30
20 - 30	25	15
30 - 40	35	12
40 - 50	45	8
50 - 60	55	6
60 - 70	65	20

By using first method:

Range (R) = L – S
= 70 – 10
= 60 Ans.

$$\text{Co-efficient of Range} = \frac{L - S}{L + S}$$

= (70 – 10) / 70 + 10
= 60/80
= 0.75 Ans.

By using second method:

Range (R) = L – S
= 65 – 15
= 50

$$\text{Co-efficient of Range} = \frac{L - S}{L + S}$$

= (65 – 15) / (65 + 15)
= 50/80
= 0.625 Ans.

In this method of calculation of either range or co-efficient of range, the frequency are completely ignored.

(ii) Quartile Deviation

It is also called semi-inter-quartile range. It is one-half of the difference between the third quartile and the first quartile.

Symbolically

$$\text{Quartile Deviation} = \frac{Q_3 - Q_1}{2}$$

Where :

Q_3 = Third Quartile
Q_1 = First Quartile

(iii) Co-efficient of Quartile Deviation

If quartile deviation is divided by the average value of the two quartiles, a relative measure of dispersion is obtained. It is called the Co-efficient of quartile deviation.

Symbolically

$$\textit{Co-efficient of quartile deviation} = \frac{\frac{(Q_3 - Q_1)}{2}}{\frac{(Q_3 + Q_1)}{2}}$$

$$= \frac{Q_3 - Q_1}{Q_3 + Q_1}$$

(i) Series of individual observation

Example : Calculate the quartile deviation and its co-efficient from the following data.

Marks obtained in 100	84	45	50	55	30	45	65	70	75

Solution

Arrange the marks in ascending order to calculate quartiles.

Serial No.	1	2	3	4	5	6	7	8	9
Marks obtained (100)	30	45	45	50	55	65	70	75	84

The value of Q_1 $= (N + 1)/4$
$= (9 + 1)/4$
$= 10/4$
$= 2.5$th observation

That is:
Size of 2nd observation +0.5 (size of 3rd. observation—size of 2nd observation.)

or $45 + 0.5 (45 - 45)$
or $45 + 0.5 (0)$
or 45
Thus $Q_1 = 45$ Ans.

The value of Q_3 $= 3(N + 1)/4$
$= 3(9 + 1)/4$
$= 3(10)/4$
$= 30/4$
$= 7.5$th observation.

That is:
Size of 7th. observation + 0.5 (size of 8th. observation – size of 7th. observation.)

or $70 + 0.5 (75 - 70)$
or $70 + 0.5 \times 5$
or $70 + 2.5$
or 72.5

Thus $Q_3 = 72.5$

Now,
Quartile Deviation $= \dfrac{(Q_3 - Q_1)}{2}$
$= (72.5 - 45) / 2$
$= 27.5 / 2$
$= 13.75$ Ans.

Co-efficient of Quartile Deviation $= \dfrac{(Q_3 - Q_1)}{(Q_3 + Q_1)}$
$= (72.5 - 45) / (72.5 + 45)$
$= 27.5/117.5$
$= 0.234$ Ans.

(ii) Discrete Series

Example : Calculate quartile deviation and its co-efficient from the following data:

Marks	0 - 10	10 - 20	20 - 30	30 - 40	40 - 50
No. of students	10	15	13	25	20

Solution

Marks	No. of students(f)	Cummulative frequency (C)
0 - 10	10	10
10 - 20	15	25
20 - 30	13	38
30 - 40	25	63
40 - 50	20	83
	$\Sigma f = 83$	

For Q_1,

The value of N/4 = 83/4 = 20.75:
which lies in 10 - 20 class interval.

Thus :

$$Q_1 = L + \frac{\frac{N}{2} - C}{f} \times h$$

$$= 10 + \frac{(20.75 - 10)}{15} \times 10$$

$$= 10 + \frac{10.75}{3} \times 2$$

= 10 + 21.5/3
= 10 + 7.166
= 17.166
= 17.16 Ans.

For Q_3,

The value of 3N/4 = 3 × 83/4 = 249/4 = 62.55
which lies in 30 – 40 class interval

Thus :

$$Q_3 = L + \frac{\frac{3N}{4} - C}{f} \times h$$

$$= 30 + \frac{(62.25 - 38)}{25} \times 10$$

$$= 30 + \frac{24.25}{5} \times 2$$

$$= 30 + 48.5/5$$

$$= 30 + 9.7$$

$$= 39.7 \text{ Ans.}$$

Now

Quartile Deviation $= (Q_3 - Q_1)/2$
$= (39.7 - 17.16)/2$
$= 22.54/2$
$= 11.27$ Ans.

$$\text{Co-efficient of Quartile deviation} = \frac{(Q_3 - Q_1)}{(Q_3 + Q_1)}$$

= (39.7 - 17.16)/(39.7 + 17.16)
= 22.54/56.86
= 0.396 Ans.

(iv) Mean Deviation (MD)

It is also called average deviation. So it is defined as the mean of the absolute values of the deviation from measures of central tendency (either mean, median or mode). In other words, mean deviation is the arithmetic average of the deviation of the individual item of the series from a measure of their central tendency.

The calculation of the mean deviation for ungrouped distribution or individual observation series follows the following steps:

(i) Calculate either mean, median or mode (according to the question)
(ii) Find out the deviation (|d|) of each item in a group from he respective central tendency.
(iii) Add these deviation (Σ|d|) (± signs should be ignored)
(iv) Σ|d|, then should be divided by the total number of items.

Symbolically

$$M.D. = \Sigma|d|/N$$

Example : Calculate the mean deviation from the following data.

Student	1	2	3	4	5	6	7	8	9
Marks	20	30	15	35	40	25	45	30	20

Solution

Students	Marks (×)
1	20
2	30
3	15
4	35
5	40
6	25
7	45
8	30
9	20
	Σ× = 260

Mean = ΣX/N
= 260/9
= 28.88 Ans.

Students	Marks	Deviation from Mean (28.88) $\lvert d \rvert$
1	20	8.88
2	30	1.12
3	15	13.88
4	35	6.12
5	40	11.12
6	25	3.88
7	45	16.12
8	30	1.12
9	20	8.88
		$\Sigma\lvert d \rvert = 71.12$

M.D. $= \Sigma\lvert d \rvert/N$
$= 71.12/9$
$= 7.9$ Ans.

Students	Marks (Arrange in ascending order)	Deviation from median (30)
1	15	15
2	20	10
3	20	10
4	25	5
5	30	0
6	30	0
7	35	5
8	40	10
9	45	15
		$\Sigma\lvert d \rvert = 70$

Value of median $= (N + 1)/2$
$= (9 + 1)/2$
$= 10/2$
$=$ 5th observation.

Thus median = 30 Ans.

Mean Deviation (M.D.) $= \Sigma\lvert d \rvert/N$
$= 70/9$
$= 7.77$ Ans.

The calculation of the mean deviation for grouped distribution or discrete series follows the following steps.

(i) Calculate the central tendency (mean, median or mode) of the distribution (according to the question),

(ii) Find out the mid-point of each class,

(iii) Record deviation of each mid-value from the calculated central tendency,

(iv) Multiply the frequency of each class by their respective deviation ($f \times |d|$) or ($f|d|$),

(v) Add the products obtained ($\Sigma f \times |d|$) (± sign should be ignored).

Example: Calculate the mean Deviation from the following data.

Class interval	Frequency
0 - 10	10
10 - 20	12
20 - 30	15
30 - 40	5
40 - 50	8
50 - 60	4

Solution

From mean

Class Interval	Mid-value (X)	Frequency (f)	f × X (fx)	Deviation from mean (25.19) |d|	f × |d| (f|d|)	Cumulative frequency
0 - 10	5	10	50	20.19	201.90	10
10 - 20	15	12	180	10.19	122.28	22
20 - 30	25	15	375	00.19	2.85	37
30 - 40	35	5	175	9.81	49.05	42
40 - 50	45	8	360	19.81	158.48	50
50 - 60	55	4	220	29.81	119.24	54
		Σf = 54	ΣfX = 1360		Σf|d| = 653.80	

Mean = $\Sigma f x / N = 1360/54 = 25.185 = 25.19$

Mean Deviation (M.D.) $= \Sigma f|d|/N$

$= 653.80/54$

$= 12.107$

$= 12.11$ Ans.

From median

Value of N/2 = 54/2 = 27
Which lies in 20 - 30 class interval.

$$\text{Thus, median} = L + \frac{\frac{N}{2} - C}{f} \times h$$

$$= 20 + \frac{(27 - 22)}{15} \times 10$$

$$= 20 + 5/15 \times 10$$
$$= 20 + 10/3$$
$$= 20 + 3.33$$
$$= 23.33 \text{ Ans.}$$

Class interval	Mid-value (X)	Frequency (f)	Deviation from median = 23.33 |d|	f |d|
0 - 10	5	10	18.33	183.30
10 - 20	15	12	8.33	99.96
20 - 30	25	15	1.67	25.05
30 - 40	35	5	11.67	58.35
40 - 50	45	8	21.67	173.36
50 - 60	55	4	31.67	126.68
		Σf = 54		Σf|d| = 666.70

Now,

Mean Deviation (M.D.) = Σf|d|/N
= 666.7/54
= 12.346
= 12.35 Ans.

From Mode

Class interval	Mid-value (X)	Frequency (f)	Deviation from mode = 22.31 |d|	f |d|
0 -10	5	10	17.31	173.10
10 - 20	15	12	7.31	87.72
20 - 30	25	15	2.69	40.35
30 - 40	35	5	12.69	63.45
40 - 50	45	8	22.69	181.52
50 - 60	55	4	32.69	130.76
		$\Sigma f = 54$		$\Sigma f\,|d| = 676.90$

$$\text{Mode} = L + \frac{f_m - f_1}{2f_m - f_i - f_2} \times h$$

$$= 20 + \frac{15 - 12}{2 \times 15 - 12 - 5} \times 10$$

$$= 20 + \frac{3}{30 - 17} \times 10$$

$$= 20 + 30/13$$

$$= 20 + 2.307$$

$$= 22.307$$

$$= 22.31 \text{ Ans.}$$

Now,

Mean Deviation (M.D.) $= \Sigma f\,|d|/N$

$= 676.90/54$

$= 12.535$

$= 12.54$ Ans.

(v) Standard Deviation

It is most useful measure of dispersion. It is also known as root-mean-square deviation. So, it is defined as the square root of the arithmetic average of the square of the deviations measured from the mean. Thus, the calculation of standard deviation firstly needs the calculation of arithmetic mean and the deviation of various items from the arithmetic mean are squared. Secondly, the squared deviations are totalled and the sum is divided by the total number of items. Finally, calculate the square root of the product to get standard deviation. Symbolically, it is represented by the Greek letter sigma (σ).

Thus: $$\sigma = \sqrt{\frac{\Sigma d^2}{N}}$$

Where: σ = Standard deviation
Σd^2 = Sum of the square of deviation
N = Total number of items

Calculation of Standard Deviation

Method of calculation of standard deviation varies according to the nature of observations. For example, standard deviation can be calculated for series of individual observation or ungrouped data by using two methods.

(i) Direct Method No.1: Symbolically this method is represented as:

$$\sigma = \sqrt{\Sigma d^2/N}$$

(ii) Direct Method No.2: Symbolically it is represented as:

$$\sigma = \sqrt{\frac{\Sigma X^2}{N} - \frac{(\Sigma X)^2}{N}}$$

Example: Calculate the Standard Deviation for following data.
10, 11, 12, 13, 15, 16, 18, 20, 22, 23

Solution

Serial No.	X	d = (X – X)	d^2
1	10	– 6	36
2	11	– 5	25
3	12	– 4	16
4	13	– 3	9
5	15	– 1	01
6	16	0	00

7	18	2	04
8	20	4	16
9	22	6	36
10	23	7	49
	$\Sigma x = 160$		$\Sigma d^2 = 192$

Arithmetic mean = $\Sigma \times /N = 160/10 = 16$

$\Sigma X = 160$ Standard Deviation $= \sqrt{\Sigma d^2/N}$

$N = 10$ $= \sqrt{192/10}$

$= \sqrt{19.2}$

$X = 16$ $= 4.38$ Ans.

$\Sigma d^2 = 192$

Example: Calculate the standard deviation by using second method from the following data.

10, 11, 12, 13, 15, 16, 18, 20, 22, 23

Solution

S.No.	X	X^2
1	10	100
2	11	121
3	12	144
4	13	169
5	15	225
6	16	256
7	18	324
8	20	400
9	22	484
10	23	529
	$\Sigma X = 160$	$\Sigma X^2 = 2752$

Standard Deviation (σ) $= \sqrt{\dfrac{\Sigma X^2}{N} - \left(\dfrac{\Sigma X}{N}\right)^2}$

$= \sqrt{2752/10 - (160/10)^2}$

$= \sqrt{275.2 - (16)^2}$

$$= \sqrt{275.2 - 256}$$

$$= \sqrt{19.2}$$

$$= \sqrt{4.38} \text{ Ans.}$$

Standard deviation for discrete series or grouped data can be calculated by using the following method. Symbolically, it is represented as:

a) $\sigma = \sqrt{\frac{\Sigma fd^2}{N}}$

b) $\sigma = \sqrt{\frac{\Sigma fX^2}{N} - \left(\frac{\Sigma fX^2}{N}\right)}$

Example: Calculate the standard deviation from the following distribution.

X	22	22	32	35	40	41	47	51	55	62
Frequency (f)	5	7	10	15	13	9	6	14	13	8

Solution

X	f	fX	$d = X - \overline{X}$	d^2	fd^2
21	5	105	– 21	441	2205
22	7	154	– 20	400	2800
32	10	320	– 10	100	1000
35	15	525	– 07	49	735
40	13	520	– 02	4	52
41	9	369	– 01	1	9
47	6	282	5	25	150
51	14	714	9	81	1134
55	13	715	13	169	2197
62	8	496	20	400	3200
	$\Sigma f = 100$	$\Sigma fX = 4200$			$\Sigma fd^2 = 13482$

Arithmetic mean ($\bar{X}$) $= 4200/100 = 42$

Standard deviation (σ) $= \sqrt{\frac{\Sigma fd^2}{N}}$

$= \sqrt{13482/100}$

$= \sqrt{134.62}$

$= 11.61$ Ans.

From second method

$$\sigma = \sqrt{\frac{\Sigma fX^2}{N} - \left(\frac{\Sigma fX^2}{N}\right)}$$

Solution

X	f	X^2	fX	fX^2
21	5	441	105	2205
22	7	484	154	3388
32	10	1024	320	10240
35	15	1225	525	18375
40	13	1600	520	20800
41	9	1681	369	15129
47	6	2209	282	13254
51	14	2601	714	36414
55	13	3025	715	39325
62	8	3844	496	30752
	$\Sigma X = 406$	$\Sigma f = 100$	$\Sigma fX = 4200$	$\Sigma fX^2 = 189882$

Standard Deviation (σ) $= \sqrt{\frac{\Sigma fX^2}{N} - \frac{(\Sigma fX)^2}{N}}$

$= \sqrt{(189882/100) - (4200/100)^2}$

$= \sqrt{1898.82 - (42)^2}$

$= \sqrt{1898.82 - 1764}$

$= \sqrt{134.82}$

$= 11.61$ Ans.

Example: Calculate the standard deviation from the following grouped data.

Marks	No. of students
0 - 10	8
10 - 20	14
20 - 30	11
30 - 40	15
40 - 50	12

Solution

By using the deviation method

Marks	No. of stu- -dents (f)	Mide-value (X)	(fX)	Deviation $d = (X - \bar{X})$	d^2	fd^2
0 - 10	8	5	40	21.5	462.25	3698.00
10 - 20	14	15	210	11.5	132.25	1851.50
20 - 30	11	25	275	1.5	2.25	24.75
30 - 40	15	35	525	8.5	72.25	1083.75
40 - 50	12	45	540	18.5	342.25	4107.00
Total	$\Sigma f = 60$		$\Sigma fX = 1590$			$\Sigma fd^2 = 10765.00$

Arithmetic mean $(\bar{X})$ $= \Sigma fX/N = 1590/60 = 26.5$

Standard deviation (σ) $= \sqrt{\dfrac{\Sigma fd^2}{N}}$

$= \sqrt{10765/60}$

$= \sqrt{179.416}$

$= 13.39$ Ans.

From Second Method

Solution

Marks	No. of students (f)	Mid-value (X)	fX	X^2	fX^2
0 - 10	8	5	40	25	200
10 - 20	14	15	210	225	3150
20 - 30	11	25	275	625	6875
30 - 40	15	35	525	1225	18375
40 - 50	12	45	540	2025	24300
	$\Sigma f = 60$		$\Sigma fX = 1590$		$\Sigma fX^2 = 52900$

$$\text{Standard Deviation } (\sigma) = \sqrt{\frac{\Sigma fX^2}{N} - \frac{(\Sigma fx)^2}{N}}$$

$$= \sqrt{52900/60 - (1590/60)^2}$$

$$= \sqrt{5290/6 - (159/6)^2}$$

$$= \sqrt{881.66 - (26.5)^2}$$

$$= \sqrt{881.66 - 702.25}$$

$$= \sqrt{179.41}$$

$$= 13.39 \text{ Ans.}$$

(vi) Coefficient of Standard Deviation

It is also called as standard coefficient of dispersion. It is calculated by dividing the standard deviation by the arithmetic average. It is calculated to know the relative measure of dispersion. Symbolically it is represented as:

$$\text{Coefficient of Standard Deviation} = \frac{\sigma}{\overline{x}}$$

In the above example its value would be = 13.39/26.5 = 0.5 Ans.

(vii) Correlation

The term correlation indicates the relationship between two variables. It is defined as, the casual connection between two series or groups of data. In other words, correlation attempts to determine the degree of relationship between variables.

There are various types of correlation, which is defined as follows:

(i) Positive or Negative
(ii) Simple, Multiple and Partial
(iii) Linear or Non-linear

(i) *Positive or Negative*

Depending upon the nature of data, correlation may be positive or negative i.e. suppose, there are two variables, the values of first variable is increasing simultaneously, the values of second variable is also increasing or the values of first variable is decreasing simultaneously, the values of second variables is also decreasing, then the movement of the values of the variables in the same direction is known as positive correlation. On the other hand, when the movements of values of variable is different, then it is known as negative correlation e.g.

Positive Correlation

Height (in feet)	2	3	4	5	6	7
Weight (in kg.)	20	30	40	50	60	70

Negative Correlation

Height (in feet)	2	3	4	5	6	7
Weight (in kg.)	50	45	40	35	30	25

(ii) *Simple, Multiple and Partial Correlation*

In simple correlation only two variables are used i.e. height and weight of men. In multiple correlation more than two factors are used i.e. crop production, rainfall and fertilizers. In partial correlation, there is more than two factors but correlation is studied between any two factors, other factors are supposed to be constant.

(iii) *Linear and Non-Linear Correlation*

When the variation in the values of two variables are in a constant ratio, it is known as Linear Correlation. On the graph it shows a straight line. If the variation is not in a constant ratio, it shows a curve line on the graph and that is known as Non-Linear Correlation e.g.

Linear Correlation

X	5	10	15	20	25
Y	6	12	18	24	30

Non-Linear Correlation

X	5	7	12	20	25
Y	4	10	15	18	20

(viii) Coefficient of Correlation

Coefficient of correlation is calculated to know the degree of correlation between two variables. It is calculated by the help of different formula, which are as follows:

Karl Pearson's Method

According to him, coefficient of correlation is calculated by dividing the sum of the products of the corresponding deviations of the various items of the series from their respective means by the product of their standard deviation and the number of pairs of observations. Symbolically it is represented as:

$$r = \frac{\Sigma \times Y}{N \sigma_1 \sigma_2}$$

This formula is known as the Product Movement Formula of coefficient of correlation. The simplified form of the above formula is as follows:

$$r = \frac{\Sigma \times Y}{N \sigma_1 \sigma_2}$$

$$= \frac{\Sigma \times 'Y'}{\sqrt{N \frac{\Sigma x'^2}{N}} \times \sqrt{\frac{\Sigma Y'^2}{N}}}$$

$$= \frac{\Sigma XY}{\sqrt{\Sigma X'^2 \times \Sigma Y'^2}}$$

Example: From the following table calculate the coefficient of correlation.

Height of father (in inches)	65	66	67	67	68	69	71	73
Height of son (in inches)	67	68	64	68	72	70	69	70

Solution

X	Y	$X' = (X - \overline{X})$	$Y' = (Y - \overline{Y})$	X'^2	Y'^2	$X'Y'$
65	67	– 3.25	– 1.50	10.56	2.25	4.87
66	68	– 2.25	– 0.50	5.06	0.25	1.12
67	64	– 1.25	– 4.50	1.56	20.25	5.62
67	68	– 1.25	– 0.50	1.56	0.25	0.62
68	72	– 0.25	3.50	0.06	12.25	– 0.87
69	70	0.75	1.50	0.56	2.25	1.12
71	69	2.75	0.50	7.56	0.25	1.37
73	70	4.75	1.50	22.56	2.25	7.12
$\Sigma X = 546$	$\Sigma Y = 548$	0.00	0.00	$\Sigma X'^2 = 49.48$	$\Sigma Y'^2 = 40.00$	$\Sigma X'Y' = 20.97$

Mean $(\overline{X}) = \Sigma X/N$
$= 546/8$
$= 68.25$

Mean $(\overline{Y}) = \Sigma Y/N$
$= 548/8$
$= 68.5$

Now, Coefficient of Correlation (r) = $\frac{\Sigma X'Y'}{\sqrt{\Sigma X'^2 \times \Sigma Y'^2}}$

$= 20.97/\sqrt{49.48 \times 40}$

$= 20.97/\sqrt{1979.2}$

$= 20.97/44.48$

$= 0.47$ Ans.

The coefficient of correlation can also be calculated without calculating the deviation of various observations from the mean of the series. Symbolically it is represented as:

$$= \frac{\Sigma XY - \dfrac{\Sigma X \Sigma Y}{N}}{\sqrt{\left\{\Sigma X^2 - \dfrac{(\Sigma X)^2}{N}\right\}\left\{\Sigma Y^2 - \dfrac{(\Sigma Y)^2}{N}\right\}}}$$

Example: Calculate coefficient of correlation from the following data.

X	1	2	3	4	5	6	7	8	9
Y	9	8	10	12	11	13	14	16	15

Solution

Calculation of Coefficient of correlation.

X	Y	X^2	Y^2	XY
1	9	1	81	9
2	8	4	64	16
3	10	9	100	30
4	12	16	144	48
5	11	25	121	55
6	13	36	169	78
7	14	49	196	98
8	16	64	256	128
9	15	81	225	135
$\Sigma X = 45$	$\Sigma Y = 108$	$\Sigma X^2 = 285$	$\Sigma Y^2 = 1356$	$\Sigma XY = 597$

Now,

Coefficient of correlation (r)

$$= \frac{\Sigma XY - \dfrac{\Sigma X \Sigma Y}{N}}{\sqrt{\left\{\Sigma X^2 - \dfrac{(\Sigma X)^2}{N}\right\}\left\{\Sigma Y^2 - \dfrac{(\Sigma Y)^2}{N}\right\}}}$$

$$= \frac{597 - (45 \times 108)/9}{\sqrt{\{285 - (45)^2/9\}\,\{1356 - (108)^2/9\}}}$$

$$= \frac{597 - 4860/9}{\sqrt{(285 - 2025/9)\ (1356 - 11664/9)}}$$

$$= \frac{597 - 540}{\sqrt{(285 - 225)\ (1356 - 1296)}}$$

$$= \frac{57}{\sqrt{60 \times 60}}$$

= 57/60

= 0.95 Ans.

EXERCISE

Q.1 Find out the mean for the following distribution of age of students living in a hostel.

Age (in years)	16-18	19-21	22-24	25-27	28-30	31-33
Frequency	19	31	36	14	11	9

Q.2 Calculate mean and median from the following table.

Yield (in kg.)	No. of plots
over 0	150
" 30	135
" 60	114
" 90	98
" 120	57
" 150	48
" 180	07

Q.3 Find out the arithmetic mean, median and mode from the following data.

Marks below	15	30	45	60	75	90
No. of students	10	25	50	97	110	120

Q.4 Calculate the value of the median, quartile first (Q_1), quartile Third (Q_3) and decile first for the following frequency distribution.

Class	Frequency
40 - 50	11
50 - 60	21
60 - 70	20
70 - 80	36
80 - 90	62
90 - 100	30

Q.5 Calculate the mode of the following frequency distribution of 100 students of a department.

Height (in cm)	Frequency
145 - 150	10
150 - 155	20
155 - 160	25
160 - 165	15
165 - 170	10
170 - 175	15
175 - 180	5

Q.6 Find out the median, lower and upper quartiles, 6th decile and 70th percentile for the following distribution:

Marks	No. of students
0 - 10	8
10 - 20	12
20 - 30	6
30 - 40	25
40 - 50	30
50 - 60	25
60 - 70	17
70 - 80	14
80 - 90	10
90 - 100	3

Q.7 Calculate median and mode from the following data.

Class-interval	Frequency
0 - 7	10
7 - 14	14
14 - 21	11
21 - 28	6
28 - 35	18
35 - 42	17

Q.8 From the following frequency table calculate mean, median and mode.

Monthly rent (in Rs.)	No. of families paying the rent
40 - 60	5
60 - 80	7
80 - 100	13
100 - 120	14
120 - 140	20
140 - 160	15
160 - 180	14
180 - 200	8

Q.9 Calculate median, mode, 7th decile and 80th percentile from the following data.

Class	Frequency
7 - 11	9
11 - 15	15
15 - 19	13
19 - 23	17
23 - 27	19
27 - 31	12
31 - 35	14
35 - 39	11
39 - 43	7
43 - 47	5
47 - 51	6
51 - 55	2

Q.10 Rainfall for a month for a station is given below. Find out the average rainfall for the station.

Rainfall (in mm.)	No. of Days
0 - 20	2
20 - 40	4
40 - 60	5
60 - 80	9
80 - 100	2
100 - 120	3
120 - 140	2
140 - 160	3

Q.11 100 families of a village are classified according to the size of land holding. Calculate mean, median, 4th decile and 87th percentile.

Size of land holding (acres)	No. of families
4 - 6	16
6 - 8	24
8 - 10	13
10 - 12	14
12 - 14	11
14 - 16	8
16 - 18	5
18 - 20	3
20 - 22	4
22 - 24	2

Q.12 Find out the mean, median and mode of wages of a group of 60 workers of a mill.

Wages/Day (in Rs.)	No. of workers
20 - 24	13
24 - 28	17
28 - 32	8

32 - 36	6
36 - 40	4
40 - 44	7
44 - 48	3
48 - 52	2

Q.13 The following table gives the rainfalls at a station for 24 years. Calculate the mean, median and standard deviation.

Rainfall (in cm)	No. of years
20 - 40	2
40 - 60	3
60 - 80	6
80 - 100	4
100 - 120	2
120 - 140	3
140 - 160	2
160 - 180	1
180 - 200	1

Q.14 Find out the mean, median and mode for the distribution of daily wages of 250 workers of a factory.

Daily wages (in Rs.)	No. of workers
30 - 39	28
40 - 49	32
50 - 59	45
60 - 69	50
70 - 79	35
80 - 89	35
90 - 99	25

Q.15 Calculate standard deviation of the distribution of wages of 30 workers in a factory.

20	22	42	32	35	56
22	21	62	28	24	46
38	33	52	38	28	37
39	48	56	18	55	51
27	60	34	25	46	29

Q.16 The weekly wages of 150 workers in a factory is given below. Find out the standard deviation of weekly wages for those families.

Weekly wages	No. of workers
100 - 150	10
150 - 200	14
200 - 250	26
250 - 300	32
300 - 350	15
350 - 400	20
400 - 450	18
450 - 500	15

Q.17 Find out the standard deviation of the following frequency distribution.

Years	No. of persons
Under 10	15
" 20	32
" 30	51
" 40	78
" 50	97
" 60	109

Q.18 Calculate the mean deviation from (i) the mean, (ii) the median of the following distribution.

Marks	0 - 15	15 - 30	30 - 45	45 - 60	60 - 75	75 - 90
No. of students	13	15	8	18	10	6

Q.19 Find out the standard deviation and the coefficient of variation of the following data.

Class interval	Frequency
3.00 - 3.25	6
3.25 - 3.50	19
3.50 - 3.75	35
3.75 - 4.00	44
4.00 - 4.25	47

4.25 - 4.50	29
4.50 - 4.75	15
4.75 - 5.00	5

Q.20 The annual production of rice of a region from 1981 to 1990 for 10 years is given below. Find out the standard deviation and coefficient of variation.

Year	Rice (in 1000 quintal)
1981	12.9
1982	14.1
1983	11.8
1984	12.4
1985	16.2
1986	17.3
1987	12.6
1988	14.7
1989	15.0
1990	16.8

Q.21 Calculate variance and coefficient of variation from the following data of the rainfall for 100 days for a station.

Rainfall (in mm.)	No. of days
Less than 10	8
" 20	20
" 30	40
" 40	70
" 50	90
" 60	100

Q.22 Find out the coefficient of correlation between the use of fertilizers and productivity from the following data.

Fertilizers used (in Quintals)	7	9	12	16	18	21	24	27	30	35

Productivity of land (in Quintals)	43	49	58	94	105	117	138	151	180	210

Q.23 Calculate the coefficient of correlation between the percentage of literate persons and percentage of employed persons.

Literate persons	37.7	34.3	38.6	35.2	39.2	32.9	33.4	36.3	30.2	38.7
Employed persons	31.2	29.1	31.8	28.5	32.6	24.0	26.4	28.1	23.0	32.2

Q.24 Calculate the coefficient of correlation between the values of X and Y given below.

X	74	85	92	65	55	75	64	57
Y	121	133	152	108	103	132	119	104

Q.25 Calculate the coefficient of correlation from the following table.

X	75	81	70	76	77	81	84	75
Y	62	68	65	60	69	72	76	72

ANSWERS

(Q.1) 22.85 years

(Q.2) Mean = 106.8 kgs
Median = 106.83 kgs

(Q.3) Mean = 47.87
Median = 48.19
Mode = 51.05

(Q.4) Q_1 = 66.5
Median = 80.32
Q_3 = 87.58
D_1 = 53.33

(Q.5) 156.67 cms.

(Q.6) Median = 48
Q_1 = 34.6

	Q_3	= 63.8
	D_6	= 53.6
	P_{70}	= 59.6
(Q.7)	Mean	= 127.08
	Mode	= 130.91
	Median	= 129.00
(Q.8)	Median	= 24.50
	Mode	= 34.46
(Q.9)	Median	= 25.31
	Mode	= 23.89
	D_7	= 32.71
	P_{80}	= 36.82
(Q.10)	74.67	
(Q.11)	Mean	= 10.28
	Median	= 9.54
	D_4	= 7.00
	P_{87}	= 16.4
(Q.12)	Mean	= 30.93
	Median	= 28.00
	Mode	= 25.23
(Q.13)	Mean	= 96.17
	Median	= 70.00
	S.D.	= 46.18
(Q.14)	Mean	= 63.98
	Median	= 63.50
	Mode	= 62.00
(Q.15)	13.4	
(Q.16)	101.72	
(Q.17)	15.48	

(Q.18) (i) 20.88
(ii) 20.95

(Q.19) S.D. = 0.3997
C.V. = 10.063

(Q.20) Mean = 14380
S.D. = 1850
C.V. = 12.86

(Q.21) Variance (σ^2) = 192.155
C.V. = 43.02

(Q.22) 0.996

(Q.23) 0.96

(Q.24) 0.935

(Q.25) 0.623

10

REPRESENTATION OF STATISTICAL DATA

Statistical data provide raw materials for drawing, diagrams, diagrammatic maps, graphs, cartograms and distribution maps. Method of representations of statistical data depend upon the nature of data i.e. climatic statistics, like temperature, pressure, rainfall etc. is represented by isotherm, isobar, isohyet respectively. While agricultural and other statics may be represented by dots, symbols, shading etc. Broadly, representation of statistical data may be classified under two headings.

1. Diagram or Graphs.
2. Distributional maps.

Diagram or Graphs

Broadly the statistical diagrams can be classified into three groups.

(a) One dimensional e.g. bars, line diagrams etc.
(b) Two dimensional e.g. square, rectangular, circular diagrams etc.
(c) Three dimensional e.g. spherical, cube diagrams etc.

10.1 ONE DIMENSIONAL DIAGRAM

(i) Line Graph

Graphs are generally drawn to show relationship between two variables. These variables are off generally different in nature because one of them plotted along X-axis (horizontal axis) and another one along Y-axis (vertical axis) and trend of utilization is represented by a series of connected straight lines or by smooth curve.

Example: Draw a line graph to show the decadal population growth in India.

Years	Total population	Round numbers in crores
1901	238, 396, 327	2.4
1911	252, 093, 390	2.5
1921	251, 321, 313	2.5
1931	278, 977, 238	2.8
1941	318, 660, 580	3.2
1951	361, 088, 090	3.6
1961	439, 234, 771	4.4
1971	548, 159, 652	5.5
1981	683, 329, 097	6.8
1991	843, 930, 861	8.4

Hints

Firstly take the round figure of the population as shown in the above table. Choose a vertical scale for showing the population and horizontal scale for showing the variable of time (Fig. 10.1).

(ii) Poly or Multiple Graph

Polygraph is such type of graph in which several sets of values can be shown with distinct line. It is most suitable for showing comparative or casual relationship.

Example: Draw a polygraph to show the decadal variation of urban population by size class—India, 1901-1991

Years	Class I	Class II	Class III	Class IV	Class V	Class VI
1901	6, 661, 100	2, 892, 094	4, 006, 520	5, 334, 969	5, 158, 901	1, 562, 467
1911	7, 029, 991	2, 687, 541	4, 195, 318	5, 047, 376	4, 940, 574	1, 679, 399
1921	8, 224, 720	2, 878, 025	4, 409, 381	5, 063, 414	5, 170, 671	1 ,945, 095
1931	10, 289, 597	3, 842, 082	5, 540, 657	5, 935, 421	5, 651, 501	1, 716, 760
1941	16, 651, 745	4, 975, 530	7, 123, 533	6, 872, 240	6, 566, 116	1, 369, 501
1951	27, 506, 519	6, 136, 734	9, 688, 132	8, 398, 871	7, 992, 923	1, 906, 467
1961	39, 882, 780	8, 712, 780	13, 138, 936	9, 902, 236	5, 328, 762	596, 504
1971	61, 225, 798	11, 679, 940	17, 125, 811	11, 706, 334	4, 755, 659	472, 992
1981	94, 503, 745	18, 189, 728	22, 409, 119	14, 930, 267	5, 603, 015	783, 894
1991	138, 801, 945	23, 309, 167	28, 079, 305	16, 531, 064	5, 532, 033	613,823

Solution

Percentage distribution of urban Population by size class-India, 1901-1991.

Year	Class I Town	Class II Town	Class III Town	Class IV Town	Class V Town	Class VI Town
1901	26.00	11.29	15.64	20.83	20.14	6.10
1911	27.48	10.51	16.40	19.73	19.31	6.57
1921	29.10	10.39	15.92	18.29	18.67	7.03
1931	31.20	11.65	16.80	18.00	17.14	5.21
1941	38.23	11.42	16.35	15.78	15.08	3.14
1951	44.63	9.96	15.72	13.63	12.97	3.09
1961	51.42	11.23	16.94	12.77	6.87	0.77
1971	57.24	10.92	16.01	10.94	4.45	0.44
1981	60.42	11.63	14.33	9.54	3.58	0.50
1991	65.20	10.95	13.19	7.77	2.60	0.29

Hints

Firstly take the round figure or the percentage of the population of the individual size class as shown in the above table. Choose a vertical scale for showing the percentage of urban population and horizontal scale for showing the variable of time (Fig. 10.2).

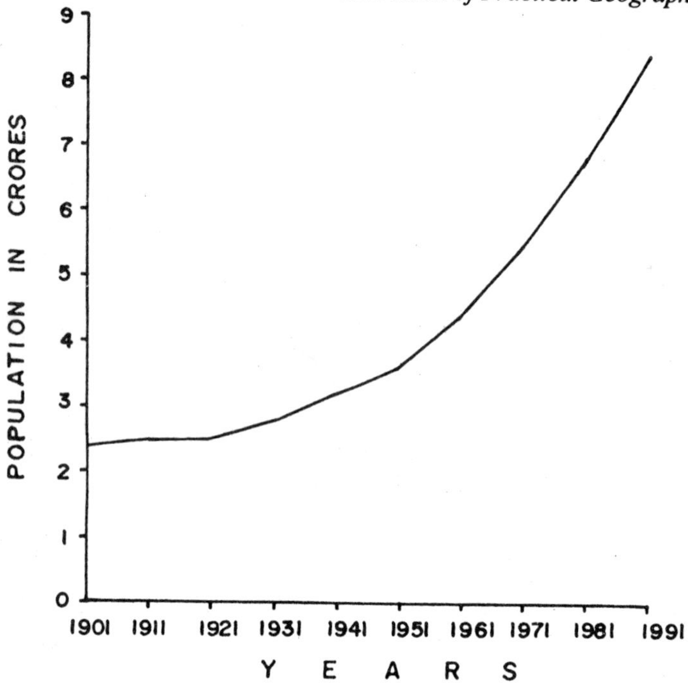

FIG. 10.1: Line Graph

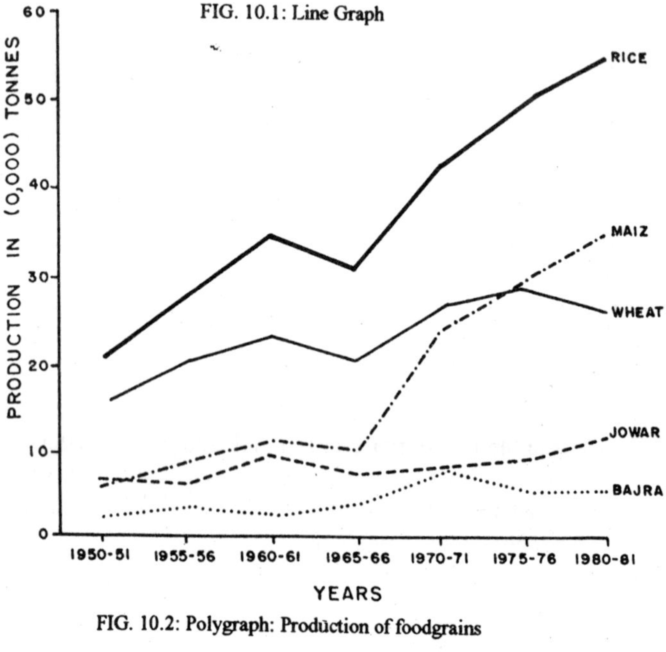

FIG. 10.2: Polygraph: Production of foodgrains

(iii) Band Graph

A band graph is virtually a line graph subdivided to show the components there of, and consequently consist of several bands formed by subdivisions and shaded in different tints (Fig. 10.3). When the variations in components are not too divergent, the band graph gives a good impression regarding their trends but when the fluctuations are too wide, its legibility is marred.

Example : Represent the following data with a suitable cartographic technique.

Years	Rice	Wheat	Jawar	Maize	Bajra
1974 - 75	39	24	5	10	3
1975 - 76	48	28	9	7	5
1976 - 77	41	29	10	6	5
1977 - 78	52	31	12	5	4
1978 - 79	53	35	11	6	5
1979 - 80	42	31	11	5	3
1980 - 81	53	36	10	6	5
1981 - 82	53	37	11	6	5

(iv) Bar Graph

It is drawn when time is fixed and the unit area varies. It is represented by bars of equal width and proportionate lengths. The unit area is represented along the horizontal axis and time or periods along the vertical axis by means of bars or columns. Bar graph are quite suitable for comparison purposes. Generally there are two types of bar graph.

1. Simple Bar Graph
2. Compound Bar Graph

Simple bar graph is used to show only one item (Fig. 10.4) but when there are several items, compound bar graph is used (Fig. 10.5).

Example: Prepare a simple bar graph with the help of following data.

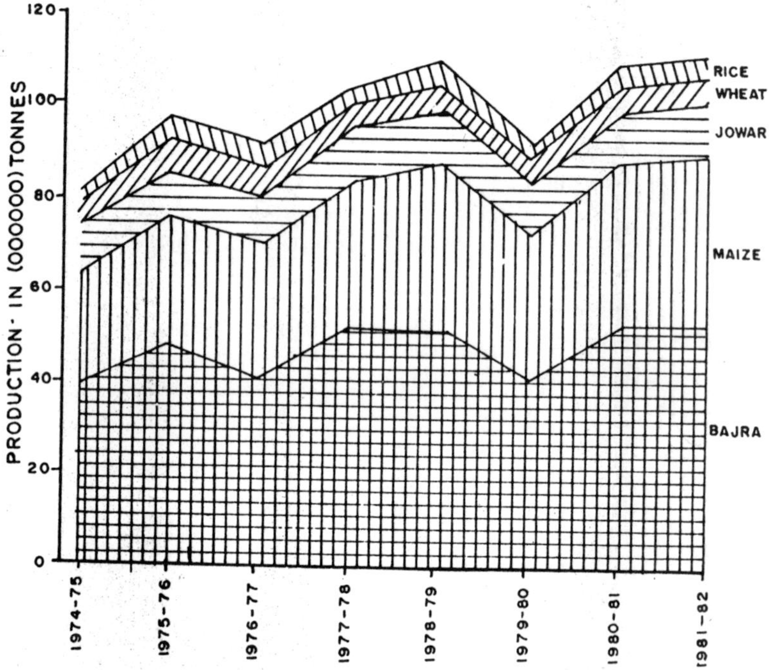

FIG. 10.3: **Band graph**: Production of Foodgrains

Year	Ratio
1901	972
1911	964
1921	955
1931	950
1941	945
1951	946
1961	941
1971	930
1981	935
1991	929

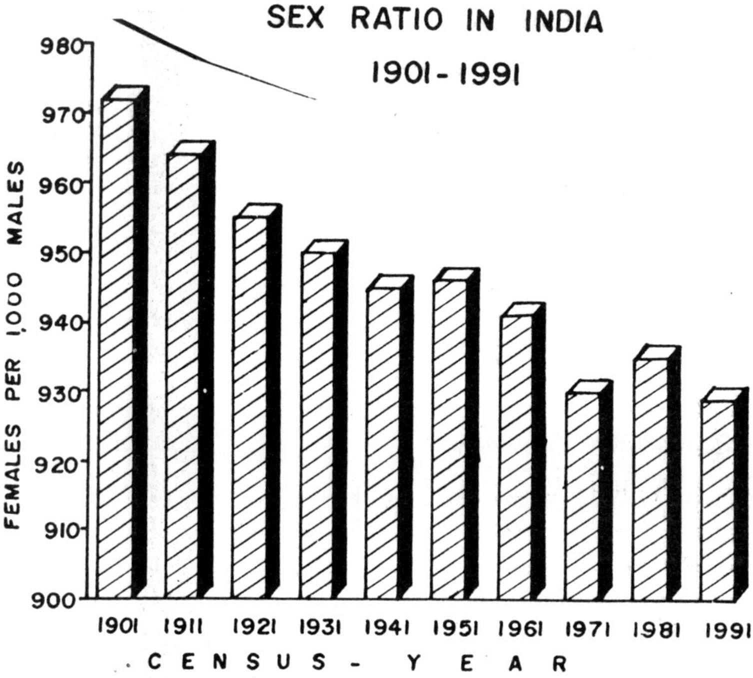

FIG. 10.4: Simple Bar diagram

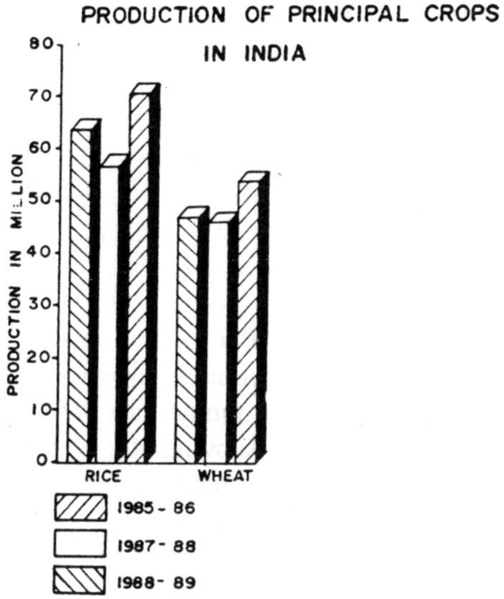

FIG. 10.5: Compound Bar diagram

Example: Construct a compound bar graph to represent the following data.

Number of urban towns by size class India, 1951-1991.

Year	I Class	II Class	III Class	IV Class	V Class	VI Class
1951	76	91	327	608	1124	569
1961	102	129	437	719	711	172
1971	148	173	558	827	623	147
1981	216	270	738	1053	739	229
1991	296	341	927	1135	725	185

(v) Multiple Bar Diagram

For two or more than two unit area Multiple bar graphs are used for comparison purposes (Fig. 10.6). If there are both positive as well as negative value, then Duo-directional bar graph is used to show the comparison (Fig. 10.6 and 10.7).

Example

(vi) Climograph or Climogram

This type of diagram was first introduced by Griffith Taylor in the second decade of 20th century, though the idea was first conceived by J. Ball.

A climograph, which represents two climatic elements or variables at any one place, one as the abscissae and the other as the ordinates. The two variables are plotted against each other. The resultant shape (twelve sided) of the graph indicates the climatic conditions of that particular place. It is most commonly used to demonstrate the relationship between mean monthly wet-bulb temperature and mean monthly relative humidity. The graph represents wet-bulb temperatures as ordinates and relative humidity as abscissae. Each of the twelve points showing wet-bulb temperature and relative humidity for individual months is marked by initial letters of the month and finally systematically joined by the help of scale. The four quadrant N.W., N.E., S.E. and S.W. represent the climatic condition as Scorching, Muggy, Raw and Keen respectively. They are very useful for comparative studies in climates.

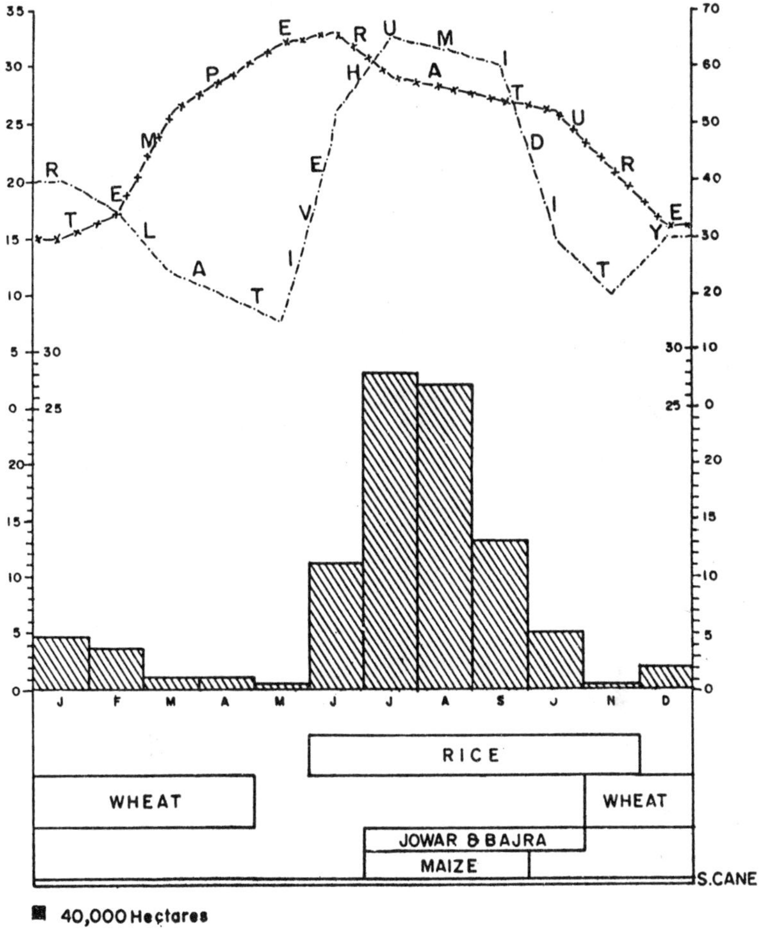

FIG. 10.6 and 10.7: Multi Bar graph

(a) *Scorching*

It represents the north-west quadrant, where throughout the year wet-bulb temperature is high varies from 70°F to 80°F and relative humidity is low less than 40 per cent.

(b) *Muggy*

It covers the area of north-south quadrant, where wet-bulb temperature as well as relative humidity is high throughout the year. The observed temperature varies from 70° F to 80° F and relative humidity always be more than 70 per cent.

(c) *Raw*

The south eastern quadrant is represented by raw, where the temperature is less than 40°F and relative humidity is more than 70 per cent.

(d) *Keen*

It represents the south-west quadrant. The observation shows low temperature and low humidity that is less than 40°F and 40 per cent respectively.

Example: Draw climographs of Calcutta and Madras with the help of following data.

Month		J	F	M	A	M	J	J	A	S	O	N	D
Calcutta	Wet bulb Temp. (°F)	64.8	68.5	70.5	78.1	82.9	82.3	80.6	80.7	80.0	78.0	68.9	66.4
	R. Humidity (%)	40	44	38	38	57	69	81	79	75	72	48	48
Madras	Wet bulb Temp. (°F)	71.3	72.1	75.0	78.8	80.8	76.6	76.7	77.0	78.0	78.5	75.5	72.5
	R. Humidity (%)	56	51	51	61	47	34	44	44	41	69	71	91

Select suitable scales to represent relative humidity along X-axis and wet-bulb temperature along the Y-axis. Draw the coordinates of relative humidity and wet-bulb temperature to mark the points of twelve months and lastly join them by the help of scale. Finally write down scorching, Muggy, Raw and Keen in the extreme corners of their respective quadrants. For better understanding sea Figs. 10.8 and 10.9.

(vii) Hythergraph

Hythergraph was also devised by G. Taylor. It is also twelve sided figure. This type of graph is represented by taking the mean monthly temperature and rainfall on Y-axis and X-axis respectively for a particular place. It is drawn in the same way in which climograph is drawn. It is drawn to compare the two elements of climate (rainfall and temperature) of different months. These two elements are important factors for the cultivation which ultimately affect the crops.

Example: Draw hythergraph of Agra and Allahabad to represent the following data.

Month		J	F	M	A	M	J	J	A	S	O	N	D
Agra	Temp. in (°F)	59.6	63.8	69.2	81.8	94.7	90.8	86.2	85.2	82.6	78.0	65.0	57.5
	Rainfall in cms.	11	10	10	8	8	19	38	35	20	15	11	8
Allahabad	Temp. in (°F)	59.5	64.9	76.8	87.6	92.5	90.8	84.5	83.2	83.0	77.6	60.5	51.8
	Rainfall in cms.	12	11	10	9	10	21	40	38	25	16	11	13

(viii) Ergograph

The Ergograph as declared firstly by A. Geddes and used by A.G. Ogilive is an important diagramatic representation. It signifies the relationship between season, climate, and its elements and crops. This type of graph shows many variables i.e. rainfall, temperature, relative humidity and acreage of various crops. The mean monthly temperature and relative humidity are represented by line graph and the mean monthly rainfall is represented by bar diagram along the vertical axis (Y-axis), the horizontal axis (X-axis) shows twelve months. Below the horizontal axis, prime base line, the acreage of various crops are shown on some selected scale.

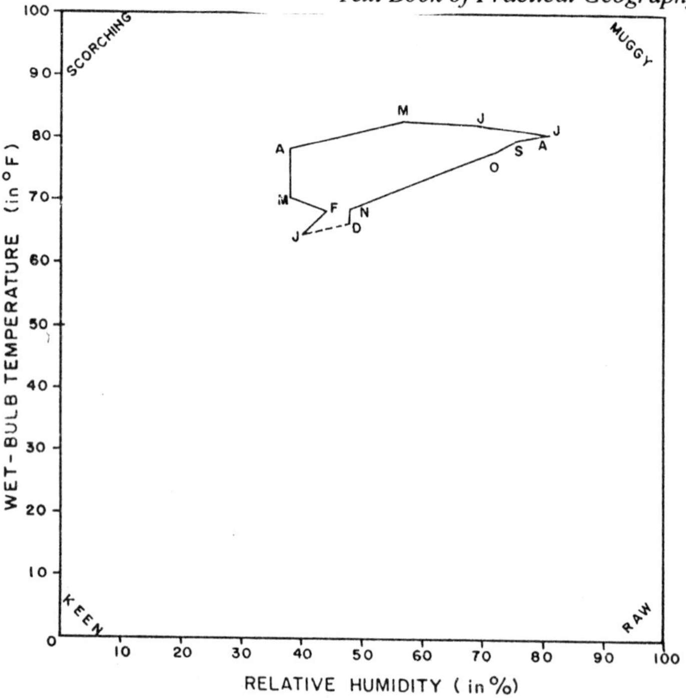

FIG. 10.8: Climographs of Calcutta

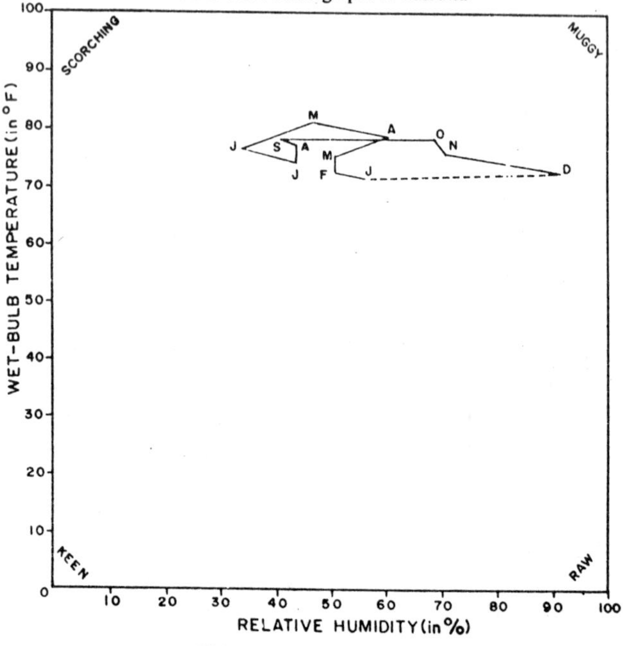

FIG. 10.9: Climograph of Madras

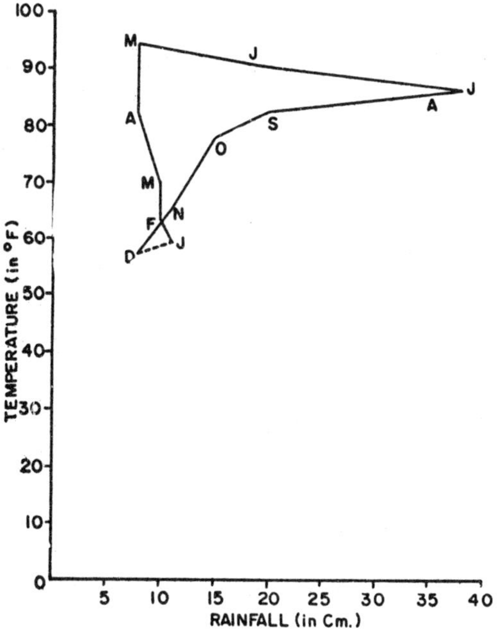

FIG. 10.10: Hythergraph of Agra

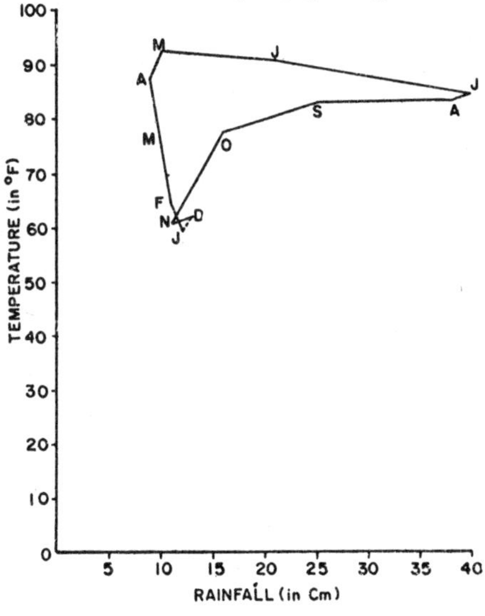

FIG. 10.11: Hythergraph of Allahabad

Example: Prepare an ergograph of Uttar Pradesh with the help of the following data.

Month	J	F	M	A	M	J	J	A	S	O	N	D
Temp. in of (°C)	15	17	26	29	32	33	29	28	27	26	21	16
R. Humidity (%)	40	35	24	20	15	52	65	63	60	30	20	30
Rainfall in cms.	4.5	3.5	1	1	0.5	11	28	27	13	5	0.5	2

The net average of various crops and their growing season.

Crops	Growing season			Net acreage (in Lakh Hect.)
Rice	June	To	November	42
Wheat	November	To	April	54
Jowar & Bajra	July	To	October	24
Maize	July	To	September	15
S.Cane	January	To	December	12
Cotton	June	To	December	7

Suppose X'X and X'Y be the two axis. Temperature, rainfall and relative humidity will be shown X'Y and XY axis and twelve months from January to December along X'X. Plot the points for each element and draw the graphs. Below X'X axis show the net hectarage of individual crop by separate rectangular block. The length of the block should be drawn on the basis of growing season and breadth should be on the selected scale for net acereage (see Fig. 10.12).

(ix) Population Pyramid

The Pyramid diagram is virtually a variant of a bar graph, where columns, constructed to represent specific qualitative population data and are arranged in a tier structure. There are generally two types of pyramid e.g. *simple* and *compound.* Simple pyramid diagram is used to show the growth of total population of a country, on the other hand a compound pyramid diagram is used to show the rural and urban components of the population or occupational structure or age and sex composition etc. For better understanding see Fig. 10.13.

Example: Construct an age and sex pyramid from the following data.

Age Group	% of Male	% of Female
0 - 9	14.6	13.3
10 - 19	10.6	10.0
20 - 29	7.8	7.6
30 - 39	6.8	6.5
40 - 49	5.2	5.0
50 - 59	3.7	3.2
60 - 69	2.1	2.2
70 +	0.2	0.2

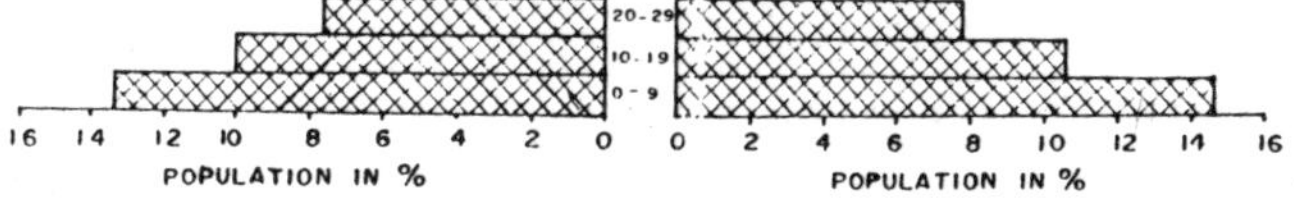

FIG. 10.13: Age and Sex pyramid

Hints

For the sake of convenience convert the population into percentage. Select a suitable scale to mark these percentage on the horizontal plane. Simultaneously mark the age group on the vertical plane in such a way that the base representing the youngest group, the apex the oldest (Fig. 10.14).

(x) Star Diagram

It is also known as wind rose or direction diagram. This type of diagram is drawn to show the direction as well as the period of movement of wind. The calm days are shown in centre by circle, which is drawn on the basis of selected scale. Different directions of the wind movement is shown by line on the same scale, which is initiated from the perimeter of the circle. The end point of the lines or rays should be jointed to give a star like appearance. So the name star diagram is given. As the rays shows the direction, hence the name direction diagram. As there are eight directions, eight radiating rays may be drawn from the perimeter of the calm day's circle (Fig. 10.15).

Example: Draw a star diagram on the basis of the following data:

No. of Days	Wind Direction	Length of wind rose in cm. (according to scale)
54	N	2.7
26	NE	1.3
24	E	1.2
38	SE	1.9
58	S	2.9
64	SW	3.2
28	W	1.4
56	NW	2.8
17	Calm	0.9

Scale 1 cm represents 20 days.

Here 1 cm representing 10 days, scale is selected. On the basis of this scale the length of the rays as well as the diameter of the circle is calculated. Finally the figure is drawn on the basis of the calculated data (Fig. 10.15).

(xi) Traffic Flow Cartogram

Traffic flow cartogram has been drawn to identify the intensity or

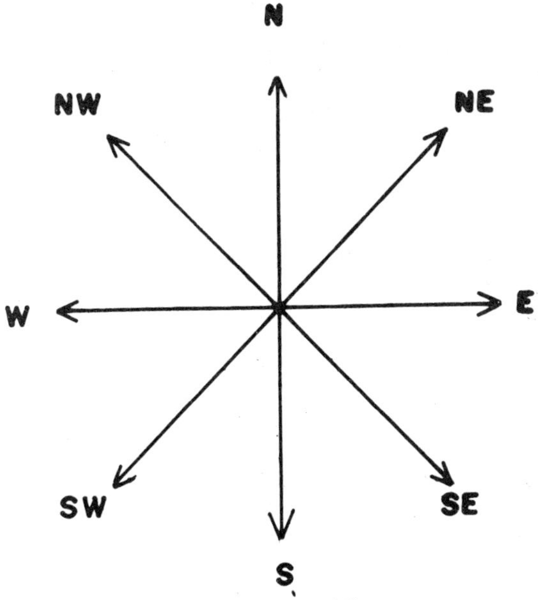

FIG. 10.14: Directional diagram

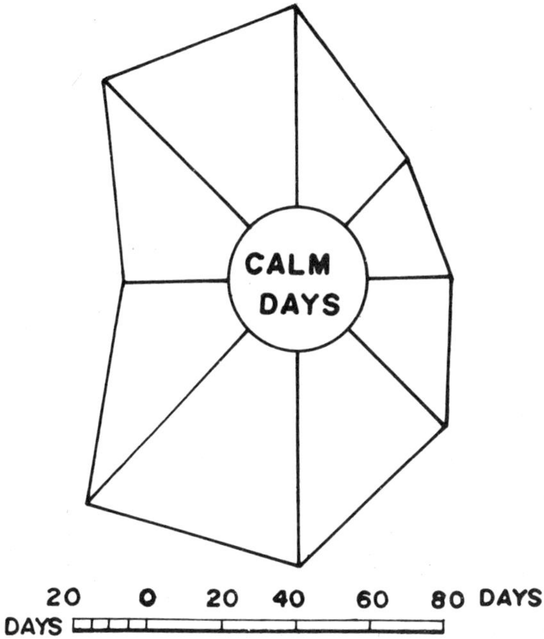

FIG. 10.15: Star diagram

concentration of traffic, goods or passengers along the lines of communication. It is drawn either by assuming proportional thickness or by a series of thin parallel lines representing proportional numbers of traffic. Due to paucity of goods and passengers data, its diagramatic representation is very difficult (Fig. 10.16).

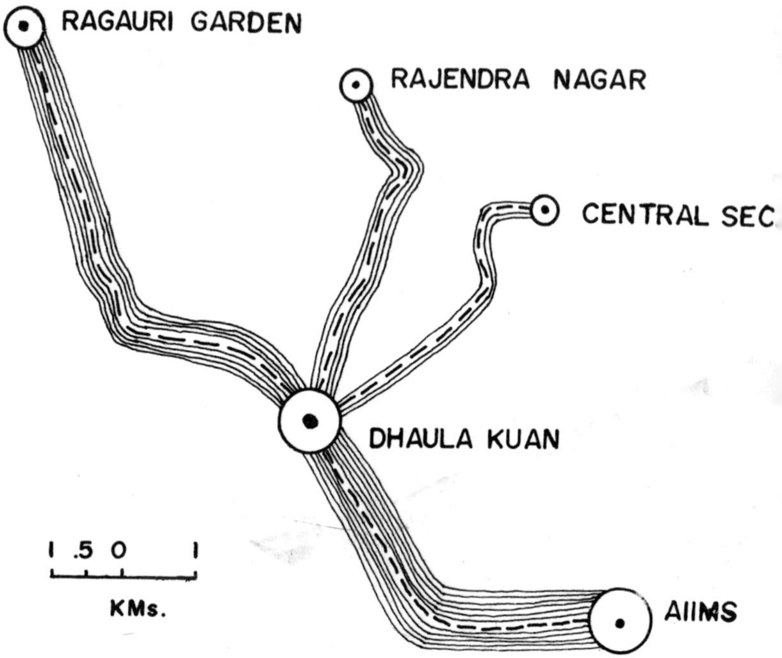

FIG. 10.16: Traffic flow diagram

10.2 TWO DIMENSIONAL OR AREAL DIAGRAM

(i) Rectangle

A rectangle is a two dimensional diagram i.e. height as well as width are taken into consideration for the purposes of representation. Rectangles are used when it is desired to give a more detailed information than can be conveyed by the 'bar'.

Example: Represent the following data by rectangle method. Also indicate the 'profit' and 'loss' per unit.

Factory	Wages Rs.	Materials Rs.	Other costs Rs.	Profit Rs.	No. of Units Produced Rs.
A	3, 000	5, 000	1, 000	1, 000	1, 000
B	2, 000	3, 000	800	500	700

The product of factory A release Rs. 10 per unit.

$$= \frac{3000 + 5000 + 1000 + 1000}{1000}$$

= 10,000/1000
= Rs. 10

The product of factory B release Rs. 9 per unit.

$$= \frac{2000 + 3000 + 800 + 500}{700}$$

= 6300/700
= Rs. 9

Thus the height of the two rectangles will be in ratio of 10:9 and their width in the ratio of units produced by the two factories i.e. 10:7.

Assuming a scale of 1 cm. = Rs. 2 and 1 cm. = 500 units, we get the diagram (Fig. 10.17).

It will be noted from the Fig. 10.18 that it is representing:

— Price per unit giving details of expenses incurred in producing each such unit;
— Profit per unit;
— Quantity sold as well as total expenses under each head.

A comparative study of budget can best be conducted by means of rectangular representation.

(ii) Square

Squares are used when it is desired to compare quantities that differ widely in magnitude. The side of a square varies as the square root of its area.

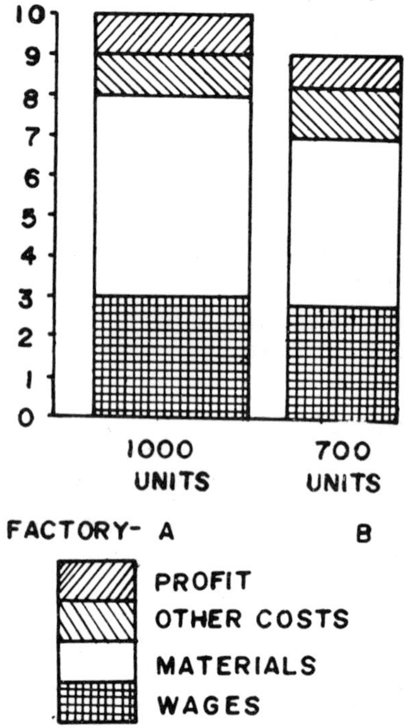

FIG. 10.17: Rectangle diagram

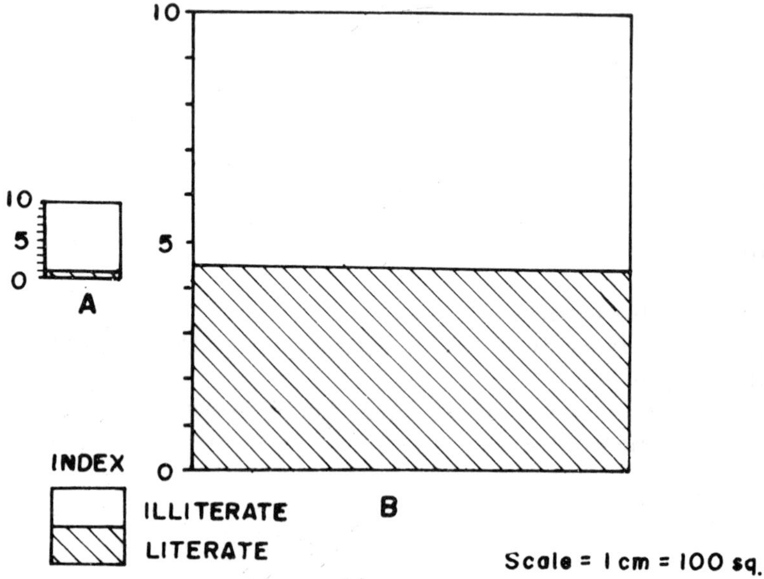

FIG. 10.18: Square diagram

Example: Represent the following data by square method.

Town	Literate	Illiterate	Total pop
A	1,000	9,000	10,000
B	1,60,000	2,00,000	3,60,000

Procedures

(a) Find out the square root of the population of the town A and B. The square roots are:
    10,000 = 100 and
    3,60,000 = 600

(b) Divide the resultant figures by some assumed figure, say 100. The quotient obtained will be 1 and 6.

(c) Draw two lines whose lengths are 1 cm. and 6 cm. respectively.

(d) Draw squares on the two lines.

(e) In order to show the literates and illiterates within their respective squares firstly find out the ratio between them.

That is:
In town A = 1 : 9
B = 4 : 5

Divide the square drawn for town *'A'* into ten (10) equal parts, and for town *'B'* into nine (9) equal parts. After that divide the square according to the ratio among the literates and illiterates and shade for identification as in Fig. 10.18.

(iii) Proportional Circle

The alternative choice of square diagrams are circular diagrams. Areas of circles vary as the squares of their radii. The formula for the calculation of radii is as:

$$r = N\sqrt{x}\,0.564$$

$$N = \pi r^2$$

$$r = \sqrt{N/\pi} = \sqrt{N/(22/7)} = \sqrt{N \text{ X } 0.564}$$

Where:
r = radii
N = Total area
0.564 = Constant

By using this formula, calculate the radii of the different circle and select a scale for convincible drawing (Fig. 10.19).

Example: Draw the proportional circle from the following data.

State	Area under Tobacco (Acres)
Maharashtra	165,951
U.P.	40,387
Bihar	35,155
Punjab	6,129

Solution

Maharashtra: $r = \sqrt{165,951} \times 0.564$
$= 407.37 \times 0.564$
$= 229.75$

U.P.: $r = \sqrt{40,387} \times 0.564$
$= 200.96 \times 0.564$
$= 113.34$

Bihar: $r = \sqrt{35,155} \times 0.564$
$= 187.49 \times 0.564$
$= 105.74$

Punjab: $r = \sqrt{6,129} \times 0.564$
$= 78.28 \times 0.564$
$= 44.14$

Suppose 100 is represented by 1 cm.
Thus:

State	Area	Radii (in cm.)
Maharashtra	165,951	2.29
U.P.	40,387	1.13
Bihar	35,155	1.05
Punjab	6,129	0.44

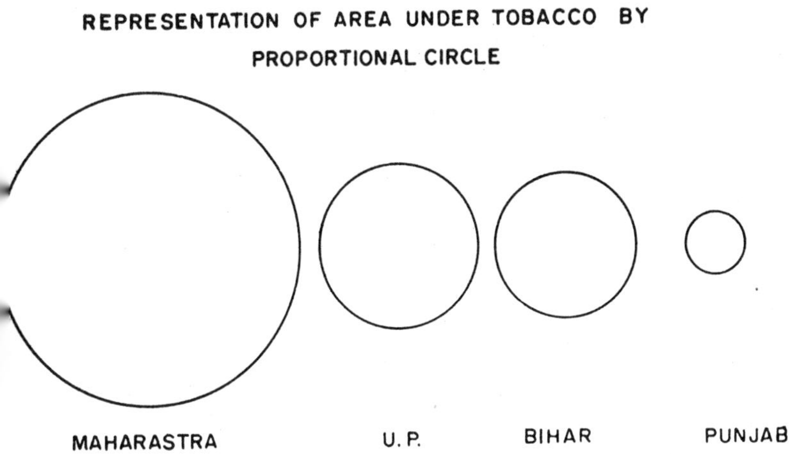

FIG. 10.19: Proportional Circle

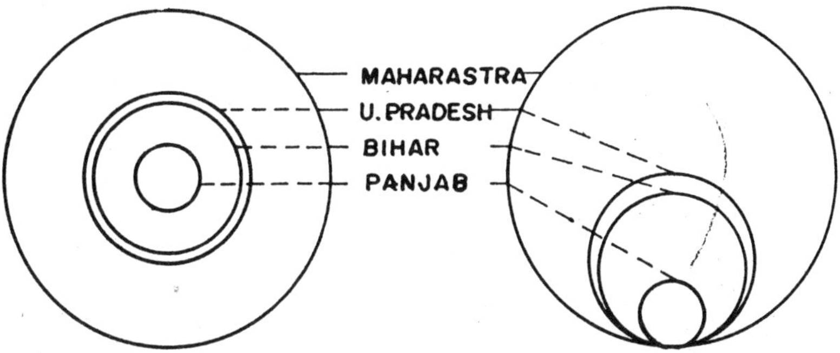

FIG. 10.20: Ring diagram

(iv) Ring Diagram

The nature of ring diagram is almost the same as proportional circle. But, there is a little difference, that is, ring diagram has a common centre for various circles instead of being drawn separately. The procedure for calculation is the same as proportional circles. But drawing is little different (Fig. 10.20).

Example: Draw the ring diagram from the following data.

S No.	State	Area under Tobacco (Acres)
1.	Maharashtra	165,951
2.	U.P.	40,387
3.	Bihar	35,155
4.	Punjab	6,129

Solution

See the example of proportional circle.

Suppose 100 is represented by 1 cm.

Thus:

S. No.	State	Area under Tobacco (Acres)	Radii (in cm.)
1.	Maharashtra	165,951	2.29
2.	U.P.	40,387	1.13
3.	Bihar	35,155	1.05
4.	Punjab	6,129	0.44

(v) Divided Circle

It is also known as Pie, Coin or Wheel diagram. In this type of diagram, the circle is divided into sectors by calculating each share in degrees. The calculation is lengthy and tedious. For better understanding see the following example.

Example: Draw a divided circle to show the proportion of production of woollen goods in the following countries.

Countries	Production [Million sq.mts. (1987)]
Italy	445.0
China	415.7
Japan	167.0
U.S.A.	154.0
Poland	143.0
India	140.0

Solution

Hints

Find out the degrees for all countries as given below.

Countries	Production [Million sq.mts. (1987)]	Degrees
Italy	445.0	109.33
China	415.7	102.13
Japan	167.0	41.03
U.S.A	154.0	37.98
Poland	143.0	35.13
India	140.0	34.40
	1465.3	360.00

Italy: 445.0/1465.3 x 360 = 109.329 = 109.33
China: 415.7/1465.3 x 360 = 102.130 = 102.13
Japan: 167.0/1465.3 x 360 = 41.029 = 41.03
U.S.A: 154.0/1465.3 x 360 = 37.982 = 37.98
Poland: 143.0/1465.3 x 360 = 35.132 = 35.13
India: 140.0/1465.3 x 360 = 34.395 = 34.40

Draw a circle on any suitable radius and divide into sectors according to the degrees calculated. Generally, the marking of degrees in a circle starts from the top of the circle either in ascending or descending order (Fig. 10.21).

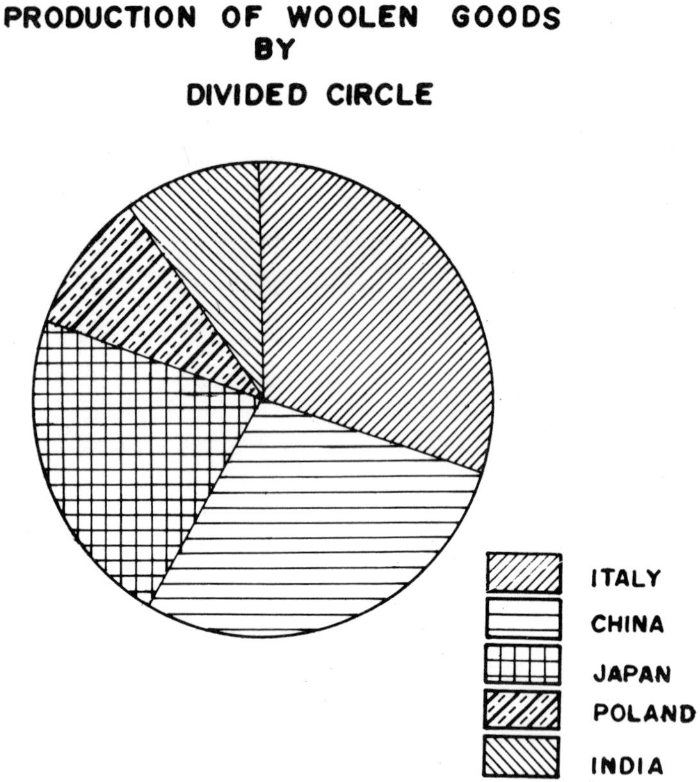

FIG. 10.21: Divided circle or Pie/wheel diagram

(vi) Proportional Divided Circle

In this type of diagram, the comparison are to be made by drawing various circles proportionate to each other and divide it into sectors (Fig. 10.22).

Example: Draw proportional divided circle to show urban population of following States by size class of town.

S. No.	States	Total urban pop. (1987)	Percentage of total population				
			Class I Town	Class II Town	Class III Town	Class IV Town	Class V Town
1.	Bihar	8,699,013	54.12	14.38	19.26	10.04	2.20
2.	U.P.	19,973,223	51.49	12.71	12.34	13.35	10.11
3.	W.Bengal	14,433,486	76.84	10.78	7.71	3.46	1.21

Solution

Hints

Firstly find out the radius of proportional circles for all the destricts from the formula as $r = \sqrt{N} \times 0.564$. After that find out the degrees for the individual circles as given below.

$r/1000$

1. $r = \sqrt{8,699,013} \times 0.564 = 1663.46 = 1.66$ cm.

2. $r = \sqrt{19,973,223} \times 0.564 = 2520.59 = 2.52$ cm.

3. $r = \sqrt{14,433,486} \times 0.564 = 2142.71 = 2.14$ cm.

Suppose 1000 is represented by 1 cm.
Now calculate the degrees as:

For Bihar

Class I Town = 54.12/100 x 360 = 194.83
Class II Town = 14.38/100 x 360 = 51.77
Class III Town = 19.26/100 x 360 = 69.34
Class IV Town = 10.04/100 x 360 = 36.14
Class V Town = 2.20/100 x 360 = 7.92

360.00

For U.P.

Class I Town = 51.49/100 x 360 = 185.36
Class II Town = 12.71/100 x 360 = 45.76
Class III Town = 12.34/100 x 360 = 44.42
Class IV Town = 13.35/100 x 360 = 48.06
Class V Town = 10.11/100 x 360 = 36.40

360.00

For W. Bengal

Class I Town = 76.84/100 x 360 = 276.62
Class II Town = 10.78/100 x 360 = 38.80

Class III Town = 7.71/100 x 360 = 27.76
Class IV Town = 3.46/100 x 360 = 12.46
Class V Town = 1.21/100 x 360 = 4.36

360.00

Thus :

S. No.	State	Radius of circle	Percentage of total population in degrees					
			Class I Town	Class II Town	Class III Town	Class IV Town	Class V Town	Total
1.	Bihar	1.66	194.83	51.77	69.34	36.14	7.92	360.00
2.	U.P.	2.52	185.36	45.76	44.42	48.06	36.40	360.00
3	W. Bengal	2.14	276.62	38.80	27.76	12.46	4.36	360.00

With the help of above table draw proportional divided circles (Fig. 10.22).

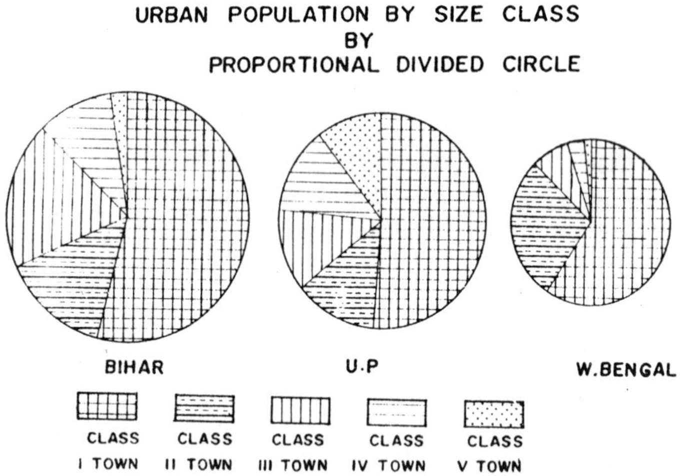

FIG. 10.22: Proportional divided circles

10.3 THREE DIMENSIONAL DIAGRAMS

Three dimensional diagrams are suitable for the large magnitude data because they help in the economy of space. It is also preferred, if the data varies too much. In this type of diagram length, width and height have to be considered. Therefore, it is also known as volume diagram. *Cube* and *spherical* diagrams are commonly used in this group. To know the side of the cube, take the cube-roots of the given data and for the calculation of radii for sphere use the following formula.

$$r = 3\sqrt{N} \times 0.62 \quad N = \frac{4}{3} \pi r^3$$

$$r^3 = \frac{3 \times N}{4 \times \pi}$$

$$= \frac{3 \times N}{4 \times \frac{22}{7}}$$

$$= \frac{3\,N \times 7}{4 \times 22}$$

$$= \frac{21\,N}{88}$$

$$r = 3\sqrt{N \times 21/88}$$

$$= 3\sqrt{N \times 0.62}$$

Where:
r = radii of the sphere;
N = total area; and
0.62 = constant

(i) Cube Diagram

Example : Draw a cube diagram to show the production of following crops in India.

Crops	Production (in Lakh tons)
Wheat	496
Cotton	114
Maize	94

Hints

Calculate the cube root of the given data and then divide them by the selected linear scale to get the side of the cube. Here linear scale is 1cm = 3. (Fig., 10.23)

Procedures

(a) Find out the cube-root of production of wheat, cotton and Maize. The cube-roots are:
(b) Divide the resultant figures by some selected figure, say 3. The quoti.

(c) Draw three lines whose length is 2.6, 1.6 and 1.5 cms. respectively.

(d) Finally draw cube on these lines.

Thus :

Crops	Cube of the total production	Side of the cube(s)	Side of the cube in cm (S/3)
Wheat	$\sqrt[3]{496}$	7.92	2.6
Cotton	$\sqrt[3]{114}$	4.85	1.6
Maize	$\sqrt[3]{94}$	4.55	1.5

CUBE DIAGRAM

PRODUCTION OF CROPS

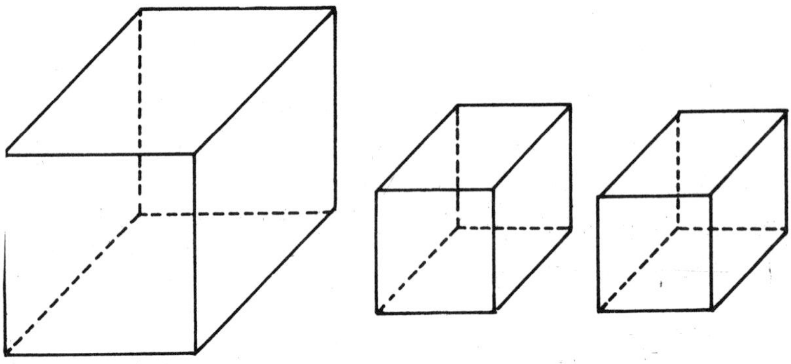

FIG. 10.23: Cube Diagram

(ii) Spherical Diagram

Example: Draw the sphere from the following data.

Language mainly spoken in households

Language	Population
Marathi	49,624,847
Kashmiri	3,174,684
Punjabi	18,588,400

Solution

Marathi: $r = 3\sqrt{N \times 0.62}$

$= 3\sqrt{49{,}642{,}847} \times 0.62$

$= 367.52 \times 0.62$

$= 227.86$

Kashmiri: $r = 3\sqrt{3,\ 174,\ 684} \times 0.62$

$= 146.97 \times 0.62$

$= 91.12$

Punjabi: $r = 3\sqrt{18,\ 588,\ 400} \times 0.62$

$= 264.89 \times 0.62$

$= 164.23$

Suppose 100 is represented by 1cm.

Thus :

Language	Population	Radii in cms.
Marathi	49,642,847	2.3
Kashmiri	3,174,684	0.9
Punjabi	18,588,400	1.6

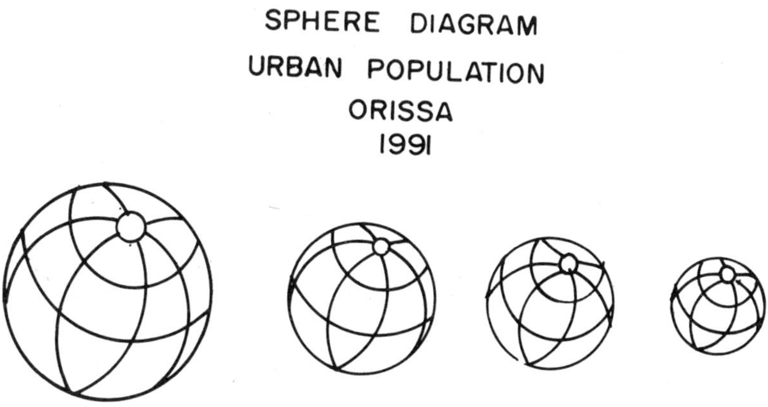

FIG. 10.24: Sphere Diagram

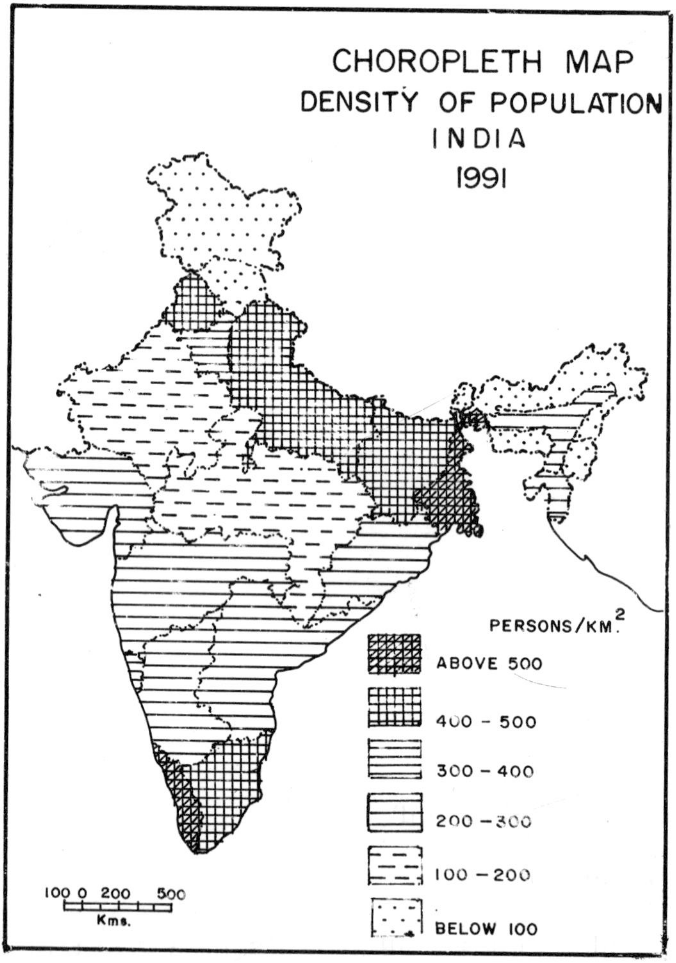

FIG. 10.25: Coropleth Map

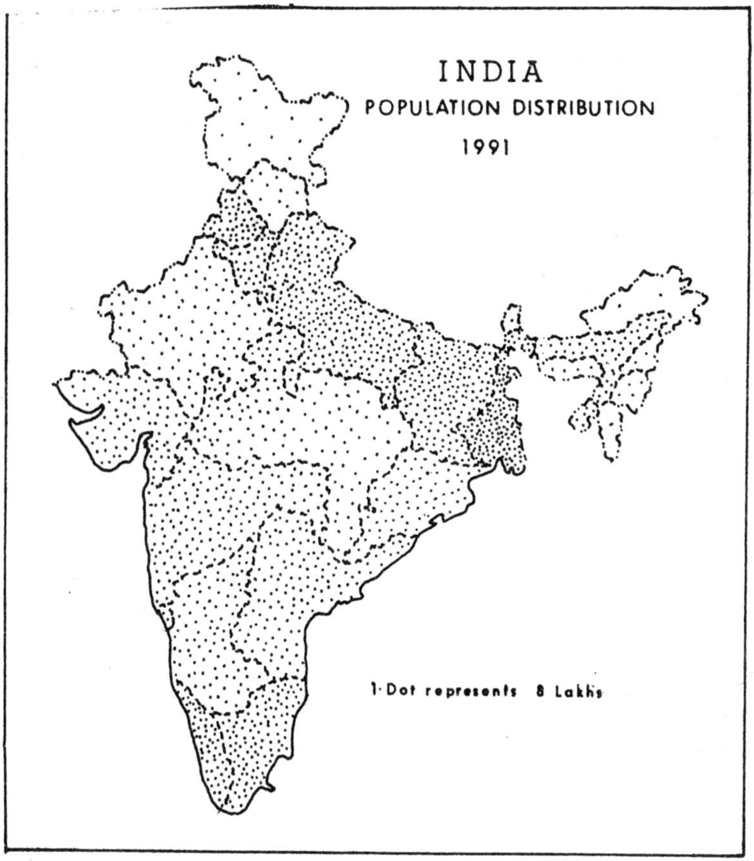

FIG. 10.26: Dot Map

10.4 DISTRIBUTIONAL MAP

(i) *Choropleth Map*

A map in which the area, density patterns are shown by using colours or shades are called choropleth maps. To provide a desired visual effect, shading and colours must be carefully chosen e.g. the distribution of average under wheat in Punjab. This distribution is available for various districts or divisions. By using suitable colours or shades, this differentiation can be shown for various districts. (Fig. 10.25)

Example : Draw a choropleth map to show the statewise density of population in India.

(ii) *Isopleth Map*

Isopleth are lines of equal measure of value or density. Isopleth join places with the same value for a certain element. All the Isopleths are counted from a datum plane which is generally zero. So a map in which quantitative distribution is shown by means of Isopleths is called an Isopleth map. Take for example, the distribution of rainfall. The Isopleth lines will show area of heavy rainfall on the one hand and areas of scarce rainfall on the other. (Fig. 10.27).

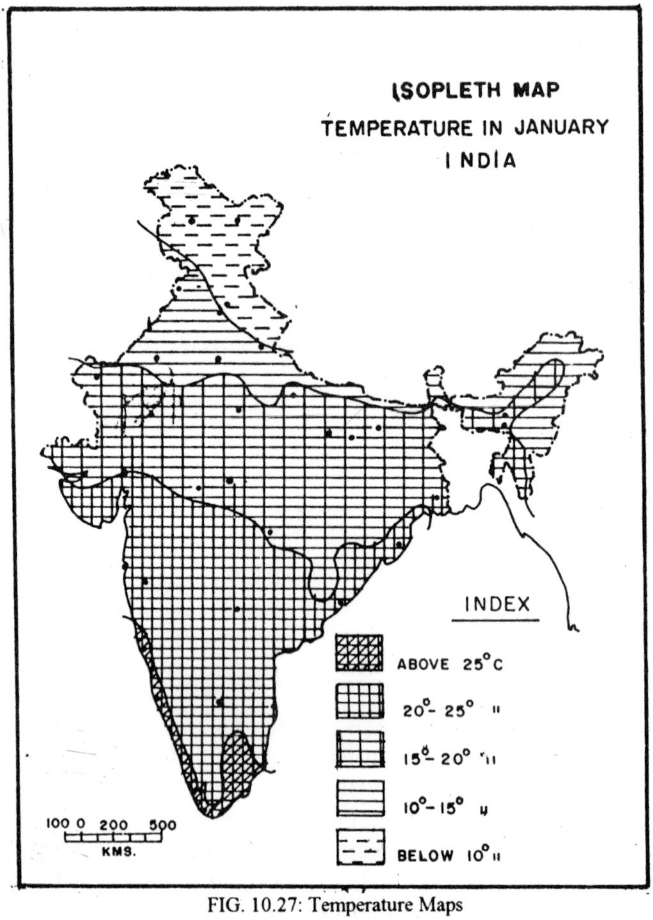

FIG. 10.27: Temperature Maps

	States	Total Pop.	Pop. in Round Figure	No. of Dots	Dots in Round Fig.	Pop. Density
1.	A.P.	66304854	66300000	132.6	133	241
2.	Arunachal Pradesh	858392	8580000	1.7	2	10
3.	Assam	22294562	22200000	44.4	44	284
4.	Bihar	86338853	86300000	172.6	173	497
5.	Goa	1168622	1160000	2.3	2	316
6.	Gujarat	41174060	41100000	82.2	82	210
7.	Haryana	16317715	16300000	32.6	33	369
8.	H.P.	5111079	5110000	10.22	10	92
9.	J&K	7718700	7710000	15.42	15	N.A.
10.	Karnataka	44817398	44800000	89.6	90	234
11.	Kerala	29011237	29000000	58	58	747
12.	M.P.	66135862	66100000	132.2	132	149
13.	Maharashtra	78706719	78700000	157.4	157	256
14.	Manipur	1826714	1820000	3.64	4	82
15.	Meghalaya	1760626	1760000	3.52	4	78
16.	Mizoram	686217	686000	1.3	1	33
17.	Nagaland	1215573	1210000	2.4	2	73
18.	Orissa	31512070	31500000	63	63	202
19.	Punjab	20190795	20100000	40.2	40	401
20.	Rajasthan	43880640	43800000	87.6	88	128
21.	Sikkim	403612	403000	0.8	1	57
22.	Tamil Nadu	55638318	55600000	111.2	111	428
23.	Tripura	2744827	2740000	5.48	5	262
24.	U.P.	138760417	138000000	276	276	472
25.	W. Bengal	67982730	67900000	135.8	136	766

(iii) *Dot Map*

A map showing distribution by means of dots, each dot of a uniform size representing a specific value. This device, for example, dot method is useful not only because it provides a good visual impression and a realistic distribution pattern, but also because it helps in understanding co-relations of distribution.

Example: Prepare a dot map to show the statewise population distribution in India

11

STUDY OF ROCKS AND MINERALS

11.1 MINERAL

A naturally occurring substance having a definite chemical composition and atomic structure and formed by inorganic processes, having particular physical properties of form, colour, streak, hardness, cleavage, fracture, lusture etc. About 2000 minerals are known. The majority of minerals are crystalline in structure and having a definite chemical composition. Some minerals form a series in which there is graduate placement of one element by another, e.g. the feldspar.

From a study of the minerals, it has been found convenient to classify them from different angles of vision. Thus, minerals may be classified as:

11.2 CLASSIFICATION OF MINERALS

(i) Rock forming minerals, and
(ii) Minerals of economic values.

In a similar manner, minerals can be classified again as

(i) Primary Minerals

It is formed as products of consolidation of magma.

(ii) Secondary Minerals

It is formed due to operation of secondary processes on the surface, subsequent to the consolidation of magma.

Minerals are classified, further, as (a) *essential minerals,* whose presence or absence affect the naming of the rocks (b) *accessory minerals,* which have no importance in the naming of the rock. However, the most scientific method of classification is based on the basis of the chemical composition, and so minerals are classified as:

(i) Native Element i.e. Gold, Copper, Diamond etc.,
(ii) Sulphides i.e. Pyrite, Galena etc.,
(iii) Sulpho-salts,
(iv) Halides i.e. rocks-salts, fluorite etc.,
(v) Oxides i.e. Quartz,
(vi) Organic salts,
(vii) Hydrocarbon compounds, and
(viii) Oxygen salts i.e. calcite, feldspar barite etc.

The study of minerals involves a knowledge of their crystalline form and internal atomic structure, their physical, optical and chemical properties and their mode of occurrence.

11.3 PHYSICAL PROPERTIES OF MINERALS

A systematic study of the minerals can never be complete without reference to their physical properties, since these are the characteristic and sometimes specific features of the different minerals.

(a) Colour

The colour of a mineral is produced due to either transmission or reflection of light and its selective absorption within the body of the mineral or upon its surface, as the case may be. The true colour of a mineral however is vitiated to a great extent due to the presence of minute quantities of impurities of different types i.e. quartz in the pure state should be absolutely colourless but it occurs commonly in shades of different colours. It is due to the presence of impurities of different types.

(b) Streak

The streak of a mineral is defined as the colour of its own powder and is determined ordinarily by producing a mark, with the mineral, upon the surface of an unglazed, white porcelain plate as is due with a piece of chalk upon the blackboard. The mineral, thus, forms a fine powder of its own and the colour of the same can be examined carefully and conveniently against the white background of the porcelain plate. If a mineral, on the other hand, is harder than porcelain, it would not produce a mark upon the plate. The streak of the mineral may be the same as its colour or it may differ radically from the same.

(c) Hardness

The hardness of a mineral is the property by virtue of which it offers resistance to any force which might tend to cause abrasion on its surface. Diamond is the hardest of all minerals and Talc is the softest minerals. With this concept in mind, a suitable concept of hardness of the minerals was proposed by Moh's in 1812 which is known as Moh's scale of hardness.

Moh's Scale of Hardness

1. Talc	6. Orthoclase
2. Gypsum	7. Quartz
3. Calcite	8. Topaz
4. Flourite	9. Corundum
5. Apatite	10. Diamond

In the above scale, the minerals have been arranged from 1 to 10 in order of increasing hardness. The hardness of any mineral is determined by comparing the same with that of the members of the scale mentioned above.

(d) Specific Gravity

The specific gravity of a mineral is the ratio of the weight of the mineral to that of equal volume of water can be determined conveniently by any standard method. If the weight of the mineral in air is given by W_1 and its weight in water by W_2, then $W_1—W_2$ is the weight of an equal volume of water and the specific gravity of mineral is given by the expression.

$$\text{Specific gravity} = \frac{W_1}{W_1 - W_2}$$

Moreover, this property of minerals is variable, within wide limits, depending upon the nature and amount of the impurities present in them.

(e) Cleavage

Cleavage is the property of a mineral by virtue of which it develops a tendency to break down, along a particular direction, offering plain and smooth surfaces. The resulting cracks, which are formed due to this tendency are known as cleavage-cracks. A mineral may or may not possess any cleavage. Only crystalline minerals are capable of possessing cleavage.

(f) Fracture

The fracture of a mineral may be defined as the nature of the surface produced due to fracturing of the mineral mass.

Following are the important types of fracture surfaces.

(i) Conchoidal

It involves the development of more or less smooth curved surfaces as in Quartz.

(ii) Even or Regular

When the broken surface is more or less smooth and plane is known as even fracture.

(iii) Uneven or Irregular

When the fractured surface is rough and uneven then it is called as uneven fracture.

(g) Lustre

The lustre of a mineral is the property by virtue of which it is capable of reflecting light. In other words, lustre of a mineral is the nature and amount of shine that the mineral can possibly offer due to reflection of light upon its surface.

The different degrees of lusture are classified as

(i) Splendent,
(ii) Shining,
(iii) Glistening,
(iv) Glimmering, and
(v) Dull.

The above five are arranged in order of decreasing amount of lustre, i.e. Splendent forms the highest degree and glimmering, the lowest, while the dull is suggestive of absence of lusture. Within the category of non-metallic lusture there are different varieties like *(i) adamantive (ii) Viterous (iii) resinous (iv) Pearly (v) Silky (vi) Greasy* etc.

(h) Transparency

On the basis of transparency there are three kinds of minerals.

(i) Transparent

When object can be seen through a thin plate of a mineral.

(ii) Translucent

When some light passes through a thin plate of minerals but objects are not identified.

(iii) Opaque

When light does not pass at all through the thin plate of minerals.

(i) Form

The form of minerals are said to embrace all the faces which have similar position in space in terms of axes of reference in the crystal. Crystals are defined as having solid geometric figures and have well-defined more or less plane faces which bound the solid. Most of the minerals are crystalline while a few are amorphous. Many crystalline minerals show some amount of flaky, fibrous, platy and similar other forms.

11.4 STUDY OF MINERAL IN HAND SPECIMENS

Minerals constitute integral part of rocks which occur upon or underneath the surface of the earth. Minerals are commonly met with, in the field, either within the rocks or in the form of segregated mineral deposits. A student should, therefore, be in a position to identify the common rock forming and economic minerals from a study of their physical properties alone. The different physical properties of minerals and the procedure for their determination have already been discussed. The physical properties of some of the common minerals are tabulated in the following paras for ready reference of the students.

11.5 ROCK

Rocks are aggregates of minerals and are the individual units constituting the crust of the earth. It may be consolidated (e.g. granite, limestone) or unconsolidated (e.g. clay, mud, sand).

There are various bases on which the rocks can be classified but the most fundamental and comprehensive is the classification of rocks on the basis of their origin. On the basis of origin there are three types of rocks.

11.6 TYPES OF ROCKS

1. Igneous rocks or primary rocks.
2. Sedimentary rocks or Secondary rocks or Aqueous rocks.
3. Metamorphic rocks or tertiary rocks.

(a) Igneous Rocks

Igneous rocks are those which have been formed by the cooling of the molten matter of the earth. In other words, igneous rocks are the products of consolidation of magmas. On the basis of formation igneous rocks are classified as :

(i) Intrusive and
(ii) Extrusive.

(i) *Intrusive Rock*

When an igneous rock has cooled quite slowly at a very great depth inside the earth, it is known as intrusive rock and is made of large crystals, granite is a typical example which is entirely crystalline.

(ii) *Extrusive Rock*

The igneous rock masses which were produced due to the consolidation of magma upon the surface of the earth are known as extrusive bodies. It is generally glassy and fine-crystalled. Basalt, lavaflows is the typical example.

On the basis of the place and time taken in cooling of the molten matter, the igneous rock may be classified into three groups.

(i) Plutonic Rock

It is formed deep in the earth. The molten matter is not able to reach the surface of the earth and cools down slowly at places situated at great depth. The rocks so formed are completely crystalline and crystals are quite big in size e.g. granite.

(ii) Dyke Rock

It is formed along the beds or the layers of the earth's crust. Some of the molten material that leaves the interior of the earth, is unable to reach the surface of the earth through the mouth of the volcano. It is semi-crystalline in structure.

(iii) Lava or Volcanic Rock

It is formed by the cooling of the lava throughout during volcanic eruptions.

The first two, plutonic and dike belong to the intrusive rock and the last lava belongs to the extrusive rocks category.

The whole classification would be fully clear from the following chart.

Igneous Rock
↓
Intrusive ↓ | Extrusive ↓

Intrusive:
↓ Plutonic (Crystalline)
↓ Dyke Rocks (Semi-crystalline)
↓ Lava or Volcanic Rocks(Glaxy)

Igneous rocks may be classified on the basis of chemical composition which is as follows :

		Chemical composition	Examples
(i)	Basic Rock	Less than 55% silica and more than 45% oxides	Basalt
(ii)	Acidic Rock	More than 65% silica	Granite
(iii)	Intermediate Rock	Between 55 to 65% silica	Andesite
(iv)	Ultra Basic Rock	Less than 45% silica	Peridolite

Characteristic of Igneous Rocks

(1) These are made of crystals of various shapes and sizes which are arranged unsystematically.
(2) These do not have any layers. But they possess well-developed joints
(3) These rocks are relatively hard and do not allow water to percolate through them.
(4) The igneous rocks do not contain any fossils or traces of animals or plants.

(b) Sedimentary Rock

Natural agencies like blowing wind, running water, percolating under ground water, glaciers in motion etc. are in ceaseless operation, causing continuous wear and tear of the rocks exposed to their fury. The product of such decay are subjected, under favourable conditions, to sedimentation and subsequent compaction. The resulting rock-masses, thus, formed under ordinary pressure and temperature conditions. The deposit of this material often occurs in the form of layers or stratas and therefore, they are known as sedimentary rocks or stratified rocks or secondary rocks or aquous rocks. Some of the most common rocks of this type are sandstone, limestone, laterite, coal etc.

According to the size of the constituent particles, sedimentary rocks are classified into three groups.

(i) ***Arenaceous or Sandy***

Which have more sand or grit. The particles that consitute them are angular and coarse. It is highly porous and can bear water, e.g. Sandstone, Grit stones, etc.

(ii) *Argillaceous or Clayey*

These rocks are softer and the minute clay particles predominate in them. These are impermeable and non-porous, e.g. shale.

(iii) *Rudaceous or Pebbly*

The particles that constitute them are comparatively big pebble or bolder, e.g. Conglomerate.

The general classification of sedimentary rocks can be understood from the following charts.

Sedimentary Rocks

- Inorganic
  - Mechanically formed
    - Arenaceous Rocks
    - Argillaceous Rocks
    - Rudaceous Rocks
  - Chemically formed
- Organic
  - Calcareous Rocks
  - Carbonaceous Rocks

Characteristics of Sedimentary Rocks

(1) The sedimentary rocks contain fossils. These fossils throw intersecting light on the place and time of their formation.

(2) These rocks are made of minute particles of various size and shape.

(3) These rocks consist of a number of layers or stratas horizontally arranged one over the other.

(c) Metamorphic Rocks

These rocks are the change form of igneous and aqeous rocks and whose characteristics have altered due to pressure and heat. Because of these forces the original rock gets harder and the mineral content also undergoes transformation. The process of formation of metamorphic rock is known as metamorphism.

According to the forces that change the rock, metamorphism may be said to be of the following two varieties.

(i) ***Thermal or Contact Metamorphism***

When rise in temperature is practically the only important factor that dominates the scene, the process is defined broadly as thermal metamorphism. Metamorphism suffered by the country rocks, in the neighbourhood and due to the heat of an igneous intrusive body, is known as contact metamorphism. For example limestone is transformed into marble, clay and shale into slate etc.

(ii) ***Dynamic or Regional Metamorphism***

Directed pressure can dominate at or near the surface of the earth while temperature dominates at considerable depths underneath. If, however, there occurs an igneous intrusion, in a region where mountain-building processes are in operation, either of the factors are likely to have a chance to dominate together. In such cases, the country rocks are, therefore, subjected to changes, both in mineral composition and texture and the process is defined as regional metamorphism. For example, Granite is converted into Gneiss, Clay and Shale into Shist.

11.7 STUDY OF ROCK IN HAND SPECIMEN

In any kind of rock—igneous, sedimentary or metamorphic, some of the rock forming minerals occur in abundance while others occur only in small amounts. Moreover, all rocks are characterised by the presence of typical textures of their own. The study of any rock, in hand specimen, naturally involves identification of the principal constituent minerals and the texture. A few common rock types are described in the following paras:

Igneous Rock

(i) ***Granite***

It is a completely crystalline rock without glassy matter and the texture varies from fine to coarse. Usually light coloured (white, reddish, greyish etc.), but may be medium or even dark coloured. The specific gravity is medium varies from 2.63 to 2.75. The essential minerals are feldspar and quartz. Biorite and muscorite occur in typical varieties.

(ii) ***Rhyolite***

The texture is generally perphyritic. The specific gravity is generally low, but may be medium in some cases varies form 2.30 to 2.70. It is generally light coloured (light grey, yellow, pale, red or even dark red) and has conchoidal fracture.

(iii) ***Pegmatite***

The texture is generally graphic. The specific gravity is usually medium. Quartz and feldspar are the essential minerals. Grains are extremely coarse and randomly oriented. It is off light coloured. It will, thus, be seen that a pegmatite differs from a granite only in grain size, other characters being essentially identical.

(iv) ***Syenite***

It is a completely crystalline rock and resembles granite in appearance but it contains feldspar together with little or no quartz. It is commonly light coloured varies from nearly white to light or deep grey. The specific gravity generally medium varies from 2.6 to 2.8 depending on the kind and proportion of minerals. Texturally, the rock is mostly medium grained with grains interlocking and randomly oriented (i.e. granitoid texture).

(v) ***Diorite***

It is crystalline medium or coarse grained rock. The main constituent minerals are feldspar and hornblend. Other ferromagnesiam minerals include biotite, pyroxene etc. The colour of the rock varies form shades of dark green to almost black. The specific gravity varies from medium to high (i.e. 2.85 to 3.00). Texturally, the rock is either medium or coarse grained having subhedral and nearly equal in size with grains interlocking (i.e. granitoid texture).

(vi) ***Gabbro***

It is massive even granular rocks. It consists essentially of pyroxene and feldspar. It is of dark-coloured with high specific gravity varying from 2.9 to 3.2. Texturally, it is medium or coarse grained. Grains are interlocking more or less of same size and randomly oriented.

(vii) *Dolerite*

It is finely crystalline dyke rock. Its constituents include pyroxene and feldspar. It is dark in colour with high specific gravity varying from 2.64 to 3.12. Texturally, the rock is fine or medium grained. Feldspar lattice occur in a criss-cross fashion.

(viii) *Basalt*

It is a volcanic rock. The grains are generally extreme fine. In colour the basalts vary from grey black to black and rather dull in appearance. Specific gravity is high (sometimes medium) varying from 2.9 to 3.1. Vesicular structure is generally found and the vesicles (i.e. cavities may be filled up with some secondary minerals like calcite, quartz etc (amygdaloidal structure).

(ix) *Peridotite*

It is made up entirely of ferromagnesiam minerals. Sometimes other minerals are present, which include pyroxene, amphibole etc. It is crystalline, massive and sometimes show a mottled appearance. Grains are usually coarse and interlocking. It is generally dark coloured varying from shades of green to complete black. The specific gravity is high varying from 3.00 to 3.30.

(x) *Diorite*

It is wholly crystalline rock. The main constituent minerals are plagioclose and hornblend. The colour varies from shades of dark green to black. The specific gravity is medium to high varies from 2.85 to 3.00.

(xi) *Pumice*

It is a glassy rock extremely porous in nature. The colour is generally white or grey but some darker varieties are also found.

Sedimentary Rocks

(i) *Sandstone*

The colour varies widely from grey, brown, white buff to red depending on the colour of the cementing material. The specific gravity varies from low to medium. It is a rock made up of chiefly of sand grains. Apart from sand, other constituents are quartz, feldspar, mica etc. The grains are rounded as well as angular and are cemented together by siliceous, calcareous or ferruginous material. The rock is usually massive but sometimes bedded structure may be visible.

(ii) *Shale*

The colour varies widely from grey, buff, yellow, red, brown, purple, green to black. The specific gravity is low. It is a finely stratified rock and is compacted muds, clays or silts. Apart from this mica and quartz grains are also common. Identity of the minerals can hardly be established since grains are too fine. Bedding planes being very close-spaced, the rock assumes a laminated character. Fossils are commonly found in some shales. It is easily scratched by knife. After moisturisation it gives muddy smell.

(iii) *Limestone*

The colour varies widely. It is white when pure and impure varieties varies from grey to black. The specific gravity is low to medium varies from 2.5 to 2.8. The grain size varies from a finely granular rock to a coarse fragments of shells and corals. Calcite is the chief constituent. Apart from this clay, quartz, dolomite and silica may be present. Texturally, the rock is very fine-grained and massive. Sometimes bedding planes may be well developed.

(iv) *Conglomerate and Breccia*

The colour varies and may have different shades rather than a continuous shades of colour. It is entirely of an acquous origin. Its constituents are the large fragments of various minerals and rocks, which are cemented together by siliceous, ferruginous and calcareous material. The texture presents a heterogeneous appearance. In conglomerate the shape of the fragments are rounded while in braccia they are angular.

(v) *Laterite*

The colour varies from red to brown. It consists mainly of limestone together with some fragments of quartz or other minerals. Texturally, the rock is massive and usually having a porous structure.

Metamorphic Rocks

(i) ***Quartzite***

The colour varies widely i.e. white, grey, yellowish, greenish, reddish or blackish. The specific gravity is medium. It is mainly formed by the metamorphism of sand stone. Mineralogically, it is made up mainly of quartz but a small portion of mica and feldspar may be present. It is medium grained rock and is compact. The texture is massive granular. The cleavage may also be present.

(ii) ***Slate***

The colour is generally black but grey, purple, red or green slates also occur. The specific gravity is medium and is about 2.75. It is chiefly formed by the metamorphism of shales having fine grained minerals. It is compact and is characterised by lusturous or dull cleavage planes. It is scratched by knife.

(iii) ***Marble***

The colour is commonly light i.e. white, grey, yellow, green, red etc. but black coloured is also found, which is produced from bituminous limestone. The specific gravity is medium. It is formed by the metamorphism of limestone. The chief constituent mineral is calcite which is associated with dolomite. It is massive and has generally an even grained texture.

(iv) ***Schist and Phyllite***

It is a completely crystalline rock. The grain size is variable from fine to medium. The colour depending widely from grey, brown, green to black depending on the ferromagnesium minerals i.e. mica schist having the colour grey to brown. The specific gravity is also extremely variable depending on the mineral composition. The mineral composition is any but constant. Mica associates with quartz, however, are more common. Chlorite, garnet, staurolite etc. are the other constituent in some other varieties when the rock is extremely fine grained and show a slaty cleavage but are characterised by the presence of mica and gives a silver appearance, it is called phyllite.

(v) *Gneiss*

The colour varies from white, shades of grey, red, brown, green to nearly black. The specific gravity is variable depending on the mineral composition. Lighter bands are usually made up of quartz and feldspar while darker bands consists usually of ferromagnesiam minerals like biolite, amphibole, pyroxene, garnet, kyanite etc. The size of the grains are usually medium or coarse and are interlocked. Alternate bands of different mineral composition is the most characteristic structural feature of this rock.

(vi) *Amphibolite*

The colour is dark with high specific gravity. The main constituent minerals are hornblend and feldspar. The size of the grains are usually medium or coarse and are interlocked. The structure is usually massive granular.

11.8 SOME COMMON MINERALS AND THEIR CHARACTERISTICS

1	2	3	4	5	6	7	8	9	10	11	12
Name	Colour	Hardness	Lusture	Streak	Fracture	Composition	Sp. Gr	Cleavage	Crystals	Form & Structure	Special Properties
Amphibole	Black	5-6	Vitreous, to Pearly	Uncoloured	Uneven	Silicate of Ca, Mg, & Fe mainly	2.9-3.4	2 Perfect sets with acute angle between them	—	Crystalline	—
Apatite	Usually, Sea green, Sometimes white	5	Vitreous	White	Conchoidal to uneven	Calcium phosphate with Ca, Cl_2 or Ca F_2	3.17-3.23	Imperfect	—	Massive, granular to compact	—
Azurite	Azure blue	3.5-4.0	Vitreous	Light blue	Conchoidal	2Cu Co_3 Cu $(OH)_2$	3.77-3.89	Perfect	—	Massive earthy	Effervesces with nitric acid
Barite	Commonly white or colourless	2.5-3.5	Generally Vitreous also resinous and Pearly	White	Uneven	Ba SO_4	4.3-4.6	3 Set Perfect	—	Crystalline often granular	—

(Contd.)

SOME COMMON MINERALS AND THEIR CHARACTERISTICS *(Contd.)*

1	2	3	4	5	6	7	8	9	10	11	12
Name	Colour	Hardness	Lusture	Streak	Fracture	Composition	Sp. Gr	Cleavage	Crystals	Form & Structure	Special Properties
Bauxite	Whitish, greyish to ochre yellow	Variable	Dull and earthy	White, Pale yellow or Pale brown	Uneven	Al_2O_3, $2H_20$	2.55	Absent	—	Pisolitic and Oolitic	—
Beryl	Shade of green, blue or white	7.5-8	Vitreous Subresinous	White	—	3 BeO. ALO_3 6 SiO_3	2.7	—	—	—	—
Biotite	Black	2.5-3	Pearly, also Sub-metallic	Uncoloured	—	H_2 K $(Mg, Fe)_3$ Al $(SiO_4)_3$	2.7-3.1	Perfect, One Set	—	Crystalline flaky	—
Calcite	White, Shade of Pink Colour	3	Vitreous earthly	White	—	Ca CO_3	2.72	Perfect	Large Crystals Common	Crystalline in rhombohedral or other forms	Effervesces briskly with cold & dilute HCl
Chalcedony	White, greyish, blue etc	6-7	Waxy or greasy	—	Conchoidal	SiO_2	2.60-2.64	Absent	—	Massive	—

(Contd.)

SOME COMMON MINERALS AND THEIR CHARACTERISTICS *(Contd.)*

1	2	3	4	5	6	7	8	9	10	11	12
Name	Colour	Hardness	Lusture	Streak	Fracture	Composition	Sp. Gr	Cleavage	Crystals	Form & Structure	Special Properties
Chalcopyrite	Golden yellow	3.5-4.0	Metallic	Shining greenish black	Uneven	$Cu Fe S_2$	4.1-4.3	Indistinct	—	Generally massive	—
Chromite	Brownish black to dull black	5.5	Sub-metallic	Brown	—	$FeOCr_2 O_3$	4.5-4.8	—	Angular Crystals	—	—
Cinnabar	Cochineal-red	2.0-2.5	Adamantine and metallic	Scarlet	Sub Conchoidal to uenven	HgS	8.0-8.2	Perfect	—	Crystalline granular, or as incrustation	—
Copper	Copper coloured	2.5-3	Metallic	—	Hackley	Cu	8.8	—	Cubic crystals	—	—
Corundum	Reddish, Grey or Greenish	9	Vitreous	Colourless	Conchoidal Uneven	AlO_3	3.9-4.1	—	Elongated Crystals	—	—
Cryolite	Colourless to snow-white	2.5	Greasy	White	Uneven	$Na_3 Al F_6$	2.95-3.00	Absent	—	Massive	Powdered mineral becomes invisible when placed in water

SOME COMMON MINERALS AND THEIR CHARACTERISTICS *(Contd.)*

1	2	3	4	5	6	7	8	9	10	11	12
Name	Colour	Hardness	Lusture	Streak	Fracture	Composition	Sp. Gr	Cleavage	Crystals	Form & Structure	Special Properties
Diamond	Colourless or yellow, red, orange	10	Adamantine	Colourless	—	C	3.5	Octhedral Octahedral Crystals	Transparent	—	—
Dolomite	White with light shade	3.3-4	Vitreous	—	Conchoidal	Ca CO_3, Mg CO_3	2.8-2.9	—	—	—	—
Feldspar	White shades of blue, Red Green	6-6.5	Vitreous	White	—	Na_2O Al_2O 6 SiO_2	2.6-2.74	Well developed	Elongated Crystals	—	—
Fluorite	Usually rosered sometimes green	4	Vitreous	White	—	CaF_2	3.01-3.25	3 Sets Perfect, Cubic	—	Crystalline cubic habit	—
Galena	Lead Grey	2.5-2.75	Metallic	Lead grey	—	PbS	7.4-7.6	3 Sets Perfect cubic	—	Crystalline cubic	Sometimes mark paper

SOME COMMON MINERALS AND THEIR CHARACTERISTICS *(Contd.)*

1	2	3	4	5	6	7	8	9	10	11	12
Name	Colour	Hardness	Lusture	Streak	Fracture	Composition	Sp. Gr	Cleavage	Crystals	Form & Structure	Special Properties
Garnet	Red, Pink, yellow, brown blue, etc	6.5-7.5	Vitreous to resinous	White	Uneven	Complex Silicate with Ca, Mg, Fe, Mn, Al, Cr, Ti, etc	3.15-4.30	No cleavage Parting Present	—	Massive; Sometimes in the form of dodica-hedron	—
Goethite	Blackish Brown	5.0-5.5	Imperfect adaman-tine to Submetallic	Brownish yellow to ochre yel-low	Uneven	Fe_2O_3, H_2O	4.28	Perfect	—	Massive botryoidal	—
Graphite	Iron black	1.5-2	Metallic	Black	—	C	2.15	—	Columnar	Foliaceous	Mark paper
Gypsum	Grey or	1.5-2	Metallic	White	Conchoidal	Ca SO_4,	$2H_2O$ 2.32		Distinct	Elongated	Crystalline
—	white								Crystals	Platy, Pris-matic, foli-ated also massive and granular	
Haematite	Grey to Black	5.5-6.5	Metallic Earthly Dull	Violet red Subconch-oidal	Uneven to	Fe_2O_3	4.90-5.30	—	Tabular	Massive Crystal granular	— sometimes

SOME COMMON MINERALS AND THEIR CHARACTERISTICS *(Contd.)*

1	2	3	4	5	6	7	8	9	10	11	12
Name	Colour	Hardness	Lusture	Streak	Fracture	Composition	Sp. Gr	Cleavage	Crystals	Form & Structure	Special Properties
Halite	Colourless or white, also yellowish	2.5	Vitreous	White	Conchoidal	NaCl	2.1-2.6	3 Sets Perfect Cubic	—	Crystalline or massive granular	Soluble in water, Saline taste
Hornblends	Deep Greenish black to black	5-6	Vitreous	Colourless	—	Ca, Mg Al, Fe, Na	3.3-4.7	Well developed	Granular Crystals	—	—
Ilmenite	Iron black	5-6	Submetallic	Brownish black	Conchoidal	FeO, TiO_2	4.5-5.0	1 Set Perfect	—	Crystalline Sometimes Platy also massive	Feebly magnetic
Kyanite	Bluish or greenish white	Parallel elongation 4-5, Transverse to elongation 6-7	Vitreous to pearly	White	—	Al_2O_3, SiO_2	3.56-3.67	2 Sets, very Perfect	—	Crystalline usually bladed	—
Limonite	Ochre yellow	5.0-5.	Dull and earthy	Yellowish brown	Uneven	$2Fe_2O_3$, $3H_2O$	3.6-4.0	Absent	—	Botryoidal, mammillary or stalactitic	—

SOME COMMON MINERALS AND THEIR CHARACTERISTICS *(Contd.)*

1	2	3	4	5	6	7	8	9	10	11	12
Name	Colour	Hardness	Lusture	Streak	Fracture	Composition	Sp. Gr	Cleavage	Crystals	Form & Structure	Special Properties
Magnesite	White	3.5-4.5	Vitreous	White	Conchoidal	$MgCO_3$	3.0-3.12	Perfect, 3 Sets	—	Massive earthy	Effervesces with acid on heating
Magnetite	Iron black	5.5-6.5	Submetallic	Black	—	Fe_3O_4	5.19	—	Octahedral Crystals	—	—
Malachite	Green	3.5-4.0	Vitreous dull or earthy	Light Green	Subconchoidal	$CuCO_3Cu(OH)_3$	3.9-4.03	Perfect	—	Massive botryoidal	Effervesces with acid
Mica	White or Black	2-2.5	Pearly	White	—	K_2Al_2 Mg_2 Fe	2.7-3.1	Perfect	—	—	—
Muscovite	White or shade of Pink, Green Black	2-2.5	Pearly	—	—	$K_2O.\ 3Al_2O_3\ 6SiO_2\ 2H_3O$	2.76-	Perfect	—	Crystalline Flaky	—
Olivine	Shade of Green, Brown Yellow	6-7	Vitreous	Colourless	—	$(Mg, Fe)_2SiO_4$	3.2-4.5	—	Crystaline	—	—
Opal	White, yellow red, Brown etc	5.5-6.5	Vitreous Pearly and Greasy	White	Conchoidal	SiO_2, nH_2O	1.9-2.3	Absent	—	Massive	—

SOME COMMON MINERALS AND THEIR CHARACTERISTICS *(Contd.)*

1	2	3	4	5	6	7	8	9	10	11	12
Name	Colour	Hardness	Lusture	Streak	Fracture	Composition	Sp. Gr	Cleavage	Crystals	Form & Structure	Special Properties
Orpiment	Lemon yellow	1.5-2.0	Resinous	Yellow	Uneven	As_2S_3	3.4-3.5	Perfect One Set	—	Foliated Crystalline	—
Orthoclase	Colourless White Pale Yellow or flesh red	6	Vitreous; Pearly and cleavage surfaces	White	—	$KAl\ Si_3O_8$	2.56-	2 Sets Perfect 2.58	—	Crystalline	—
Psilomelane	Iron black	5-7	Submetallic to dull	Brownish black	Sub conchoidal to uneven	MnO_2	3.3-4.7	Absent	—	Massive and botryoidal	—
Pyrite	Bronze or Brass yellow	6-6.5	Metallic	Greenish or Brownish Black	—	$Fe\ S_2$	4.8-5.1	—	Cubic Crystals	—	—
Pyrolusite	Iron black	2.0-2.5	Metallic	Black	—	MnO_2, with little water	4.73-4.86	Absent	—	Massive granular	Soil fingers
Pyroxene	Black	5-6	Vitreous	Uncoloured	Uneven	Silicate of Ca,Mg & Fe mainly	3.1-3.6	2 Perfect Sets more or less perpendicular to each other	—	Crystalline	—
Quartz	Colour-less to white	7	Vitreous	White	Conchoidal	SiO_2	2.65-	Absent 2.66	—	Commonly massive; Sometimes with crystal faces	—

Some Common Minerals and their Characteristics *(Contd.)*

1	2	3	4	5	6	7	8	9	10	11	12
Name	Colour	Hardness	Lusture	Streak	Fracture	Composition	Sp. Gr	Cleavage	Crystals	Form & Structure	Special Properties
Realgar	Aurora-red	1.5-2.0	Resinous	Orange-red	Conchoidal	As S	3.56	Perfect, One Set	—	Crystalline granular or as incrustation	—
Siderite	White	3.5-4.0	Vitreous, inclining to pearly	White	Uneven	$Fe\ CO_3$	3.83-3.88	3 Sets Perfect	—	Massive	Effervesces with cold HCl
Sphalerite	Brown, yellow or black	3.5-4.0	Resinous to adamantine	Brownish light yellow or white	—	Zn S	3.9-4.1	6 Sets Perfect	—	Crystalline	—
Stibnite	Lead grey	2	Metallic	—	Lead grey	Sb_2S_3	4.52-	Perfect 1 Set 4.62	—	Acicular or fine columnar radiating or as irregular aggregates	—
Talc	Silky white shades of green	1	Pearly	—	White	$3MgO\ 4\ SiO_2\ H_2O$	2.7-2.8	Perfect	Granular Crystals	—	—
Topaz	Shades of yellow Blue, Pink	8	Vitreous	—	Colourless	$Al_2\ F\ SiO_4$ (+ OH)	3.5-5.6	Well developed	Alongated Crystals	—	—
Tourmaline	Bluish black	7.0-7.5	Vitreous to resinous	Subconchoidal to uneven	—	—	2.98-3.20	Imperfect	—	Massive columnar crystals Prismatic in habit	—

12

ELEMENTARY AREAL PHOTOGRAPHY

The invention of Photography by the French Scientist Lowis Daguerie in 1839 was the first milestone that widened man's ability to detect things beyond his own senses, especially that of sight. The French geodesist Aacago recommended the use of photographs for preparation of topographical maps in 1840. Another Frenchman Colonel A. Laussedat may be regarded as the founder of Photogrammetry, though he called his technique 'Merophotographic'. This progress in photography led to the development of photogrammetry and photo interpretation.

12.1 PHOTOGRAMMETRY

Photogrammetry is defined as the science of obtaining reliable measurements by means of photography. That is in photogrammetry, we use photographs for various purposes such as for the construction of topographical maps, preparation of mosaics, for evaluation of forest resources, classification of soil, urban planning etc.

12.2 PHOTO INTERPRETATION

Photographic interpretation is the art of examining photographic images for the purposes of indentifying objects and judging their significance. This art or technique is extensively used by the engineers, surveyors, foresters, geologists, urban planners etc.

Photogrammetry is divided into two distinct branches.

(1) Aerial Photogrammetry
(2) Terrestrial Photogrammetry

(1) Aerial Photogrammetry

It is that branch of photogrammetry in which the measurements are made from photographs taken from an airborne camera.

(2) Terrestrial Photogrammetry

In this type of photogrammetry, the measurements are made from photographs taken from points on the ground.

Aerial photogrammetry was extensively used after 1909, that is after the aeroplane was invented by Wilbur Wright. After 1930's, the use of air photographs began to be made at commercial level. U.S., Canada, U.K., Germany etc established Air Survey Companies, which conducted aerial surveys.

12.3 TYPES OF AERIAL PHOTOGRAPHS

Generally two types of Aerial photographs are used in Air-Surveys.

(1) Vertical Photographs
(2) Oblique Photographs

(1) Vertical Photographs

Vertical photographs are those made with the camera pointed down vertically or as nearly vertically as is practical in an aircraft (Fig. 12.1).

(2) Oblique Photographs

Oblique photographs are taken with the camera axis directed intentionally between the horizontal and the vertical i.e. with the camera pointed obliquely (Fig. 12.1).

Oblique photograph are of two types

(a) High oblique is that in which the camera is pointed more obliquely and the apparent horizon is shown (Fig.-12.1).

(b) Low oblique are relatively less oblique than the high oblique photographs and the horizon is not shown in this type of oblique photographs (Fig. 12.1).

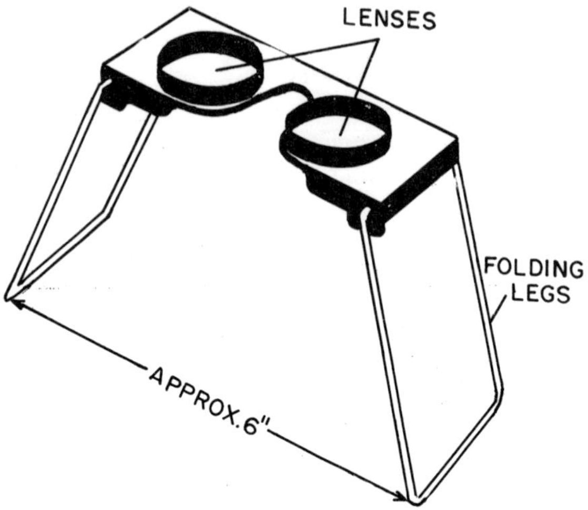

FIG. 12.1: Pocket stereoscope

Besides these general types there are three more types of Aerial photographs.

(i) Convergent photographs
(ii) Trimetrogon photographs
(iii) Terrestrial photographs

12.4 GENERAL EQUIPMENT USED IN AIR PHOTO INTERPRETATION

In this photo interpretation, we use many instruments which help us in the interpretation of these Aerial Photographs.

The generally used instruments are:-

1. Pocket Stereoscope

2. Mirror Stereoscope
3. Parallax Bar
4. Scanning Stereoscope
5. Sketch Master

1. Pocket Stereoscope

The most important parts of a pocket stereoscope are the two single positive lenses with focal length less than 250 mm, which are mounted at normal eye base (65 mm) apart. The lenses are approximately 100 mm above the plain of the table. The magnification is given by 250/100 = 2.5 times. This type of stereoscope is best suited only for small format photographs (Fig. 12.2).

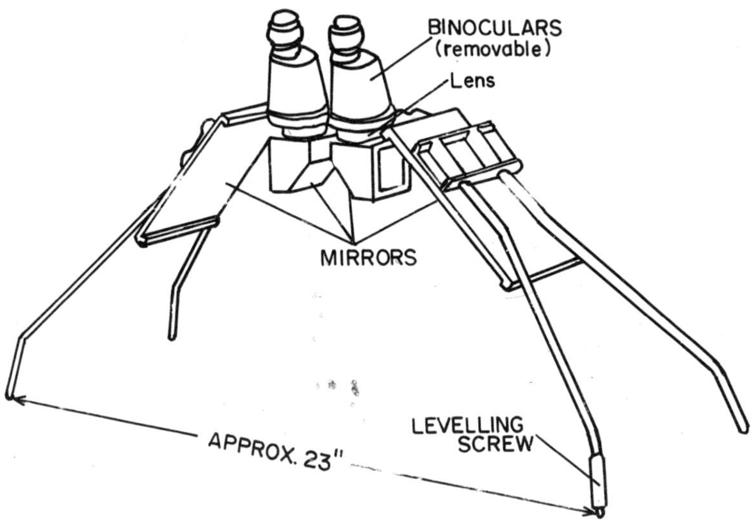

FIG. 12.2: Mirror Stereoscope

2. Mirror Stereoscope

Two or more often a pair of mirrors, a prism and a lens are the main components of a mirror stereoscope. The distance from the lens to the image plane is about 300 mm. so the refraction power of the lens is lessened and there is a reduction to 0.8 times. We also use binoculars

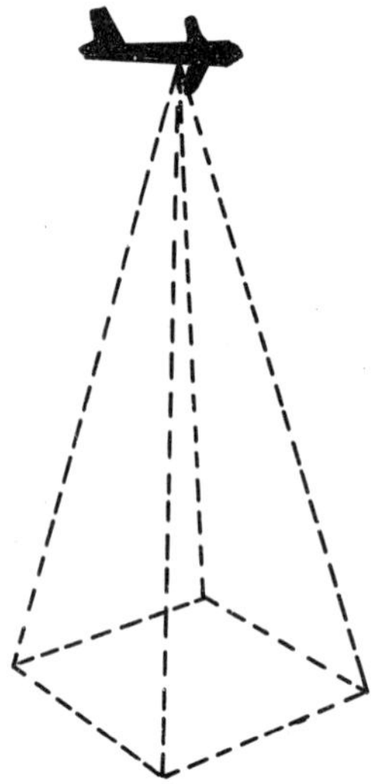

FIG. 12.3: Diagram of height determination

above the lenses to increase the occular capacity (Fig. 12.3).

3. Scanning Stereoscope

This stereoscope is similar to mirror stereoscope but it allows the operator to roam around the steremodel at low or high magnification without the need to move the aerial photographs.

4. Parallax Bar

A Parallax Bar consists of two glass plate separated by an adjustable metal bar. On each plate there is a small spot or cross and this is termed a half mark. This instrument is used to measure the height of objects on aerial photographs.

5. Sketch Master

The Sketch Master is used in mapping or in delineation of landscapic features on available topographical maps. It contains a metallic stand with graduated vertical scale fixed to any geometrically shaped metallic piece, to the stand is attached another metal piece by an adjustable screw. In this metal piece an arm is horizontally fixed, which is also graduated with scale and which can move forward or backward. On the other side of a metal piece, a metal plate is attached on which photographs are fixed with the help of magnetic metallic weights. At the end of the horizontal arm, another horizontal bar is perpendicularly attached in which a double prism is fixed.

12.5 SCALE DETERMINATION OF VERTICAL AERIAL PHOTOGRAPHS

The Scale of an aerial photograph can be determined through a number of ways. In order of accuracy, these methods are:-

(1) By establishing the relationship between the photo distances and the corresponding ground distances.
(2) By establishing the relationship between the distances in the photographs and the corresponding distances on the map.
(3) By establishing relationship between the focal length *(f)* of the camera and the Flying height *(H)* of the Aircraft.

Method - I

By establishing Relationship between photo distance and the ground distance

When the corresponding distances of an aerial photographs and of the ground are known, then the scale of the photograph can be determined by using the following equation SP = DP/DG.

Example: The distance between two points as measured on ground is 200 mts. and the corresponding distance between the images of the same two points on an aerial photographs is 1.0 cm. Then the scale of the photograph would be

Solution

$$SP = \frac{DP}{DG}$$

$$\frac{1 \text{ cm}}{200 \text{ mts.}}$$

$$\frac{1 \text{ cm}}{200 \times 100 \text{ cms.}}$$

$$\frac{1}{20{,}000}$$

Method - II

By establishing relationship between photo distance and the map distance

The scale of an aerial photograph can be determined readily, if a reliable map of the same area is available. The photo scale may then be computed by measuring the distances on the aerial photographs and the distances between the corresponding points on the map. The following relationship is used to determine the scale of photographs.

$$\frac{\text{Photo Scale}}{\text{Map Scale}} = \frac{\text{Photo distance}}{\text{Map distance}}$$

$$\text{Photo scale} = \frac{DP}{DM} \times msf$$

Where: *DP* is the distance on the photograph,
DM is the corresponding map distances,
msf is the map scale factor.

Example: The distance measured between two points on the map is found to be 2.0 cm. The distance between the corresponding images on the aerial photographs was measured as 10 cm. Find the scale of photographs, when the scale of the map is 1/250,000.

Solution

$$SP = \frac{DP}{DM} \times msf$$

$$= \frac{10}{2} \times 250,000$$

$$= \frac{1}{50,000}$$

Method - III

By Establishing Relationship Between Focal Length (f) and Flying Height (H).

In case of truly vertical photograph of a relatively flat terrain, the scale is obtained as

$$\frac{f}{h}$$

If we have got the focal length of the aerial camera and also know the flying height *(H)* of the Aircraft from the datum plane. Then scale can be easily calculated.

Example: An aerial photograph was taken from an aerial camera having focal length (f) of 6 inches and was flown at a height (h) of 10,000 feet above datum plane. The average height of the ground level is 2000 feet above datum. Calculate the average scale of aerial photograph.

Solution

$$\text{S-avg.} = \frac{f}{\text{H} - \text{H-avg.}}$$

$$= \frac{6}{(10,000 - 2,000) \times 12}$$

$$= \frac{1}{16,000}$$

Where: *S avg* is average scale of the aerial photograph
f is the focal length
h is the flying height
H avg. is the average height of the ground level above datum.

12.6 CHARACTERISTICS OF PHOTOGRAPHIC IMAGES

As the objects on the most Aerial photographs are at very small scale. There are some characteristics of every object which help and assist in its Identification. Keeping these characteristics of objects in mind, an Interpreter identifies the objects on these photographs. The following most important characteristics of photographs assist in their identification.

Size

The size of an object is one of the most useful clues to its identity. When we have to identify an unknown object, we measure its size and then comprising this size with the size of other objects on the photograph, we identify the object. Realizing that size can be three dimensional. The interpreter can make parallel measurement where necessary.

Shape

The shape of the object largely helps in the interpretation of the objects. Sometimes the shape of the objects seen in vertical photography create difficulty in interpretation. The ability to understand and make use of the plan view has to be acquired with practice. The plan view of objects is an important and sometimes conclusive indication of their structure, composition and function. The value of shape to the interpreter is that it delimits the class of objects to which an unknown must belong. It aids understanding of significance and function.

Shadow

Shadow present in aerial photograph sometimes help the interpreter by providing him with profile representations of objects of interest shadows are particularly helpful if the objects are very small or lack tonal contrast with their surroundings. Under these conditions, the sharp tonal gradients of the shadow may enable the interpreter to identify objects which themselves are just at the threshold of recognition.

Tone and Colour

Colour perception is an important element of one's awareness of one environment but in black and white photographs, distinction between hues are lost and objects are observed in tones of grey. The tones of photographic images are influenced by many factors like spectral sensitivity of an object and its background. Spectral sensitivity of the photographic film, spectral scattering by atmospheric haze particulars and spectral transmission by the photographic filter. In single photographs, where the shapes of objects must be inferred from monocular clues, tone is particularly important.

Texture

Texture in aerial photographs is created by tonal repetitions in groups of objects which are too small to be discerned as individuals. It follows that the size of object required to produce texture varies with the scale of photography. In large scale photographs, trees can be seen as individual but their leaves cannot be discerned separately but these contribute to the texture of the tree crowns. In photographs of smaller scale, the crowns contribute to the texture of the whole stand of trees.

Pattern

Pattern or the spatial arrangement of objects is an important clue in the interpretation of Aerial Photographs and this helps a lot in identifying the objects. Industries generally have a common pattern everywhere and this general pattern often leads to the correct identification of the industries on the Aerial photographs.

12.7 PROBLEMS ON AERIAL PHOTOGRAPHY

By the help of following data, find (a) the number of strips, (b) Number of photographs in each strip, (c) total number of photographs to cover the given area;

Area to be covered = 50 Kms × 30 Kms
Scale of Aerial Photograph = 1:15,000
Forward lap = 75 per cent
Side lap = 30 per cent

Solution

The effective length of the photograph perpendicular to the flight line 'L'

= 70 × 23/100
= 16.1 cm

The effective length of the photograph perpendicular to the flight line on 1:15,000 scale 'LS' = 16.1 × 15,000/100,000
= 2.4 km.

Given width of the area is 30 km.

Therefore, number of strips to cover 30 km = 30/2.4
= 12.5

No of strips (in round figure) = 13

The effective width of an aerial photograph along the flight line on 1:15,000 scale 'Ws' = 5.75 × 15,000/100,000
= 0.86 km

Therefore, number of aerial photographs to cover
50 km = 50/0.86
= 58.1

To ensure full stereoscopic coverage, add 4 more photographs per strip. Therefore the number of photographs per strip = 58 + 4 = 62

The total number of photographs per strip = 62 Ans.

The total number of photographs to cover an area of 50 km × 30 km
62 × 14 = 868 photographs

The total number of photographs required = 868 Ans.

APPENDICES

APPENDIX - I

CONVERSION TABLES

Linear			**Area**			**Capacity**		
Inches		**Centimeters**	**Sq. Yards**		**Sq. Meters**	**Pints**		**Litres**
.393701	1	2.539998	1.19599	1	.83613	1.76072	1	.56795
.787402	2	5.079996	2.39198	2	1.67226	3.52144	2	1.13590
1.181103	3	7.619994	3.58797	3	2.50839	5.28216	3	1.70385
1.574804	4	10.159992	4.78396	4	3.34453	7.04288	4	2.27180
1.968505	5	12.699990	5.97995	5	4.18065	8.80360	5	2.83975
2.362206	6	15.239988	7.17594	6	5.01678	10.56432	6	3.40770
2.755907	7	17.779986	8.37193	7	5.85291	12.32504	7	3.97565
3.149608	8	20.319984	9.56792	8	6.68904	14.08576	8	4.54360
3.543309	9	22.859982	10.76391	9	7.52517	15.84648	9	5.11155
Yards		**Meters**	**Acres**		**Hectares**	**Gallons**		**Litres**
1.093614	1	914399	2.47104	1	.40469	22009	1	4.54360
2.187228	2	1.828798	4.94208	2	.80938	.44018	2	9.08720
3.280842	3	2.743197	7.41312	3	1.21407	.66027	3	13.63080
4.374456	4	3.657596	9.88416	4	1.61876	.88036	4	18.17440
5.468070	5	4.571995	12.35520	5	2.02345	1.10045	5	22.71800
6.561684	6	5.486394	14.82624	6	2.42814	1.32054	6	27.26160
7.655298	7	6.400793	17.29728	7	2.83283	1.54063	7	31.80520
8.747912	8	7.315192	19.76832	8	3.23752	1.76072	8	36.34880
9.842526	9	8.229591	22.23936	9	3.64221	1.98081	9	40.89240
Miles		**Kilometers**	**Sq. Miles**		**Sq. Kilometers**	**Weights**		
						Ounces		**Grams**
.621372	1	1.609343	.38610	1	2.59000	.035270	1	28.349530
1.242744	2	3.218686	.77220	2	5.18000	.070540	2	56.699060
1.864116	3	4.828029	1.15830	3	7.77000	.105810	3	85.048590
2.485488	4	6.43372	1.54440	4	10.36000	.141080	4	113.398120
3.106860	5	8.046715	1.93050	5	12.95000	.176350	5	141.747650
3.728232	6	9.656058	2.31660	6	15.54000	.211620	6	170.097180
4.349604	7	11.265401	2.70270	7	18.13000	.246890	7	198.446710
4.970976	8	12.874744	3.08880	8	20.72000	.282160	8	226.796240
5.592348	9	14.484087	3.47490	9	23.31000	.317430	9	255.145770

Area			Volume			Weight		
Sq. Inches		Sq. Centimeters	Cu. Feet		Cu. Meters	Pounds		Kilograms
.15500	1	6.45163	35.31445	1	.02832	2.204620	1	.453592
.31000	2	12.90326	70.62890	2	.05664	4.409240	2	.907184
46500	3	19.35489	105.94335	3	.08496	6.613860	3	1.360776
.62000	4	25.80652	141.25780	4	.11328	8.818480	4	1.814368
.77500	5	32.25815	176.57225	5	.14160	11.023100	5	2.267960
.93000	6	38.70978	211.88670	6	.16992	13.227720	6	2.721552
1.08500	7	45.16141	247.20115	7	.19824	15.432340	7	3.175144
1.24000	8	51.61304	282.51560	8	.22656	17.636960	8	3.628736
1.39500	9	58.06467	317.83005	9	.25488	19.841580	9	4.082328

Sq. Feet		Sq. Meters	Cu. Yards		Cu. Meters	Tons*		Kilograms
10.76387	1	.09290	1.30794	1	.76456	.000984	1	1016.0470
21.52774	2	.18580	2.61588	2	1.52912	.001968	2	2032.0940
32.29161	3	.27870	3.92382	3	2.29368	.002952	3	3048.1410
43.05548	4	.37160	5.23176	4	3.05824	.003936	4	4064.1880
53.81935	5	.46450	6.53970	5	3.82280	.004920	5	5080.2350
64.58322	6	.55740	7.84764	6	4.58736	.005904	6	6096.2890
75.34709	7	.65030	9.15558	7	5.35192	.006888	7	7172.3290
86.11096	8	.74320	10.46352	8	6.11648	.007872	8	8128.3760
96.87483	9	.83610	11.77146	9	6.88104	.008856	9	9144.4230

* I metric tonne = 1,000 kg.

APPENDIX - II

SOME USEFUL WEIGHT AND MEASUREMENT UNITS

	Linear Measure		***Weights***
10 Millimeters	= 1 Centimeter	10 Milligrams	= 1 Centigram
10 Centimeters	= 1 Decimetre	10 Centigrams	= 1 Decigram
10 Decimetre	= 1 Meter	10 Decigrams	= 1 Gram
10 Meters	= 1 Decimetre	10 Grams	= 1 Decagram
10 Decimetres	= 1 Hectameter	10 Decigrams	= 1 Hectogram
10 Hectameters	= 1 Kilometer	10 Hectograms	= 1 Kilograms
		10 Kilograms	= 1 Quintal
		10 Quintals	= 1 Ton
		1000 Kilograms	= 1 Ton

	Square Measure	***Metric Equivalents Length***
100 Sq. Millimeters	= 1 Sq. Centimeter	1 Millimeter = 0.03937 Inches
100 Sq. Decimetres	= 1 Sq. Meter	1 Centimeter = 0.3937 Inches
100 Sq. Meters	= 1 Sq. Decametre	1 Meter = 39.3701 Inches
10 Sq. Decametres	= 1 Sq. Hectameter	1 Meter = 3.2808 Feet
100 Sq. Hectameter	= 1 Sq. Meter	1 Meter = 1.0936 yards
		1 Meter = 0.54681 Fathoms
		1 Kilometer = 0.62137 Miles (land)

Cubic Measure	***Area***
1000 Cu. Millimeters = 1 Cu. Centimeter	1 Sq. Millimeter = 0.00155 Sq. Inches
1000 Cu. Centimeters = 1 Cu. Decimetre	1 Sq. Centimeter = 0.1550 Sq. Inches
1000 Cu. Decimetres = 1 Cu. Meter	1 Sq. Meter = 10.7639 Sq. Feet
	1 Sq. Meter = 1.19599 Sq. Yards
	1 Hectare = 2.47105 Acres
	1 Sq. Kilometer = 247.105 Acres
	1 Sq. Kilometer = 0.3861 Sq. Miles

Linear Measure		***Avoirdupois Weight***
12 Inches	= 1 Foot	16 Drams = 1 Ounce
3 Feet	= 1 Yard	16 Ounces = 1 Pound
22 Yards	= 1 Chain	14 Pounds = 1 Quarter

10 Chains or 220 Yds = 1 Furlong

8 Furlongs = 1 Mile

28 Pounds = 1 Quarter

4 Quarters = 1 Cwt

20 Cwt's = 1 Ton (2240 lbs)

2000 Pounds = 1 Short Ton

2240 Pounds = 1 Long Ton

Square Measure

144 Sq inches = 1 Sq Foot

9 Sq Feet = 1 Sq Yard 1 litre

1210 Sq Yards = 1 Rood

4 Roods = 1 Acre (4840 Sq yds)

640 Acres = 1 Sq Mile

Volume & Capacity

1 Cu Centimetre = 0.06102 Cu inches

= 61.024 Cu inches

1 litre = 0.0353 Cu. Feet

1 litre = 0.2642 US Gallons

1 litre = 2.2200 UK Gallons

1 Hectolitre = 26.417 US Gallons

1 Hectolitre = 21.997 UK Gallons

1 Cu Metre = 35.3147 Cu Feet

1 Cu Metre = 1.30795 Cu yards

1 Cu Metre = 264.174 US Gallons

1 Cu Metre = 219.969 UK Gallons

Liquid Measure

10 Mililitres = 1 Centimeter

10 Centilitres = 1 Decilitre

10 Decilitres = 1 Litre

10 Litres = 1 Decalitre

10 Decalitres = 1 Hectolitre

10 Hectolitres = 1 Kilolitre

Cubic Measure

1728 Cu inches = 1 Cu foot

27 Cu Feet = 1 Cu yard

128 Cu Feet = 1 Cu Cord(Wood)

40 Cu Feet = 1 Ton (Shipping)

215042 Cu inches = 1 U.S. Standard Gal

1 Cu Foot = About 4/5 of a Bushel

Liquid Measure

60 Minims = 1 Fluid Drachm

8 Fluid Drachms = 1 Fluid Ounce

20 Fluid Ounces = 1 Pint

4 Gills = 1 Pint

4 Quarts = 1 Gallon

Meight (Mass)

1 Miligram = 0.01643 Grains

1 Gram = 0.03527 Ounces

1 kilogram = 2.20462 Pounds

1 Metric Quintal = 220.462 Pounds

1 Ton = 2204.64 Pounds

APPENDIX - III

RELATION AMONG ALTITUDES, PRESSURES AND TEMPERATURES

Altitude (metres)	Pressure (millimetres)	Temperature (°C.)
– 500	806.2	+18.3
0	760.0	15.0
500	716.0	11.7
1,000	674.1	8.5
1,500	634.2	5.2
2,000	596.2	+2.0
2,500	560.1	–1.2
3,000	525.8	–4.5
3,500	493.2	–7.8
4,000	462.2	– 11.0
4,500	432.9	– 14.2
5,000	405.1	– 17.5
5,500	378.7	– 20.8
6,000	353.8	– 24.0
6,500	330.2	– 27.3
7,000	307.2	- 30.5
7,500	286.8	– 33.7
8,000	266.9	– 37.0
8,500	248.1	– 40.3
9,000	230.5	– 43.5
9,500	213.8	– 46.7
10,000	198.2	– 50.3
10,500	183.4	– 53.3
11,000	169.7	– 55.0
11,500	159.9	– 55.0
12,000	145.0	– 55.0

APPENDIX – IV

INTERNATIONAL TIME (+/– INDIAN STANDARD TIME)

Asia & Pacific

Afghanistan	– 1
Australia	
Victoria	+4½
S. Australia	+4
Bangladesh	+2½
Brunei	+2½
Burma	+1
China	+2½
Fiji	+6½
Hong Kong	+2½
India	IST
Indonesia	+1½
Japan	+3½
Malaysia	+2½
Nepal	+1/4
New Zealand	+6½
Pakistan	– ½
Papua New	
Guinea	+4½
Philippines	+2½
Singapore	+2
South Korea	+3½
Sri Lanka	IST
Taiwan	+2½
Thailand	+1½

Africa

Algeria	– 5½
Angola	– 4½
Congo	– 4½
Ethiopia	– 2½
Gabon	– 4½
Gambia	– 5½
Ghana	– 5½

Other

Sweden	– 4½
Cyprus (North)	– 2½
Cyprus (South)	– 3½
Greece	– 3½
Malta	– 4½
Turkey	– 3½

Eastern Europe

Bulgaria	– 3½
Czechoslovakia	– 4½
East Germany	– 4½
Hungary	– 4½
Poland	– 4½
Romania	– 3½

Soviet Union

Moscow	– 2½
Omsk	+ ½
Vladivostok	+4½
Anadyr	+7½

North America

Canada	
Atlantic	– 9½
Central	– 11½
Eastern	– 10½
Mountain	– 12½
Newfoundland	– 9
Pacific	– 13½
Yukon	– 13½

United State

Alaska (East)	– 13½
Alaska (West)	– 16½
Central	– 11½
Eastern	– 10½
Hawaii	– 15½

Mountain	– 12½
Pacific	– 13½
Cent. & South	
Argentina	– 8½
Bahamas	– 10½
Bolivia	– 9½
Brazil	
Aere	– 10½
East	– 8½
West	– 9½
Chile	– 9½
Columbia	– 10½
Costa Rica	– 11½
Cuba	– 10½
Ecuador	– 10½
Guatemala	– 11½
Jamaica	– 10½
Kenya	– 2½
Libya	– 3½
Mauritius	– 1½
Morocco	– 5½
Mozambique	– 3½
Nigeria	– 4½
South Africa	– 3½
Sudan	– 3½
Tanzania	– 2½
Tunisia	– 4½
Uganda	– 2½
Zaire (Kinshasa)	– 4½
(Katanga)	– 3½
Zambia	– 3½
Zimbabwe	– 3½
Middle East	
Bahrain	– 2½
Egypt	– 3½
Iran	– 2
Iraq	– 2½
Israel	– 3½
Jordan	– 3½
Kuwait	– 2½
Lebanon	– 3½
Libya	– 3½
Oman	– 1½
Qatar	– 2½
Saudi Arabia	– 2½
Western Europe	
Austria	– 4½
Belgium	– 4½
Denmark	– 4½
Finland	– 3½
France	– 4½
Iceland	– 5½
Ireland	– 4½
Italy	– 4½
Luxembourg	– 4½
Netherland	– 4½
Norway	– 4½
Spain	– 4½
Switzerland	– 4½
U.K.	– 5½
West Germany	– 4½
Portugal	– 5½

INDEX

Abney's Level, 194-95
Altitudes relations, 327-28
Anemometer, 69
Anti-cyclone, 76-77
Areal diagram, 270
Areal photography, 312-22
 characteristics of, 320-21
 distances of, 317
 establishing relationship, 317-19
 instruments, 315
 photogrammetry, 312
 photo interpretation, 312
 problems of, 321-22
 types of, 313-14
Astronomical maps, 36
Atlas maps, 35

Ball, J., 260
Band graph, 257
Bangladesh, 83
Bar graph, 257
Barometer, 73
 types of, 73-74
  aneroid, 73
  barograph, 74
  mercurial, 73
Bharat Khand Reserved Forest, 60
Bihar, 59
Biweekly markets, 61
Bluff, 105
Bonn's projection, 135-38
Burma (Myanmar), 82

Cadastral survey, 160
Campbell-stokes recorder, 72
Captain Karter, 177
Celsius scale, 62
Centigrade thermometer, 63
Chain survey, 160-61
Champaran, 59
Charopleth map, 285
Cirque, 112
Climate map, 36
Climograph, 260
Clouds, 80
 classification of, 70-72
Conical projection, 129-134-35
Common mineral,
 characteristics of, 303-311
Communications, 61
 network, 34
Comparative scale, 11
Contours, 101, 103
 relief representation by, 103-118
Conversion table, 323-24
Correlation, 238-40
Cube diagram, 280
Cultural information, 58
 maps, 36
Cyclones, 76
Cyclonical orthomorphic projection, 126

Deciles and percentiles, 210-18
Diagonal scale, 23
Distribution map, 285
Doma Khand Reserved Forest, 60
Dot map, 287
Double vernier scale, 32
Drainage, 60
Dutch cartographer, 126

Eidograph, 42
Engineering survey, 160
Erograph, 263

Fahrenheit thermometer, 63
Fiord coast, 115
First World War, 78

Gall's projection, 152-54
Gandak river, 60
Geodetic survey, 160
Geological maps, 36
 criteria for,
  faults, 87
  folding, 87

unconformities, 87
exercises of, 90
Gnomonic projection, 141
Gorakhpur, 59
Gorge, 108
Graphic scale, 5

Hachures, 99
Hanging valley, 112
Hill shading, 100
Humidity, 68, 260
Hygrometer, 68
Hythergraph, 263

Indian clinometer, 192-93
Indian Weather Map, 78-80
International agreement, 52
International time (+— Indian Standard time), 328-29
Irrigation, 61
Isobaric system, 79

Kanna Nadi, 60
Karl Pearson's method, 240
Knoll, 105

Land utilisation map, 37
Levelling survey, 184
differential, 186
objects of, 186
patterns of, 184
plotting, 182-188
profiles, 188

Maps,
classification, 34
combination of, 42
enlargement and reduction, 37
camera lucia, 42
edograph, 42
pantograph, 40
photographic method, 38
similar triangular method, 40
square method, 38
meausement of 45
projection, 119-58
Bonn's projection, 135
construction of, 120-135
graphical, 132, 135, 137, 142-45, 148, 152, 156
methematical, 122, 126
choice of, 156-58
classification, 120
conical, 129, 133
cylindrical equal area, 123, 124
definitions, 119
Gall's projection, 152
gnomonic, 138
graphical, 152
Mercator's projection, 126
natural cylindrical, 121
orthographic, 147
polyconic, 138
problems of, 157-58
stereograph, 144-45
use of, 50-51
Zenithal, 141, 148-52
Marha Nadi, 60
Maximum and minimum thermometer, 65
Mean deviation, 227-32
Median, 206
Mercator, Gerardus, 126
Meter scale, 22
Military survey, 160
Minerals, 288-292
classification of, 288
physical properties, 288
cleavage, 291
form, 292
fracture, 291
hardness, 290
lustre, 291
specific gravity, 290
streak, 289
transparency, 292
Mirror stereoscope, 315
Mode, 218-20
Mollweide's projection, 157-58
Multiple bar diagram, 260

Narayani Nadi, 60, 61
Negative Vernier scale, 30
Nepal, 59

One dimensional diagrams, 253
band graph, 257
bar graph, 257
climograph, 260
erograph, 263
hythergraph, 263
line graph, 253
poly or multigraph, 254

pop: lation pyramid, 267
star diagram, 268
traffic flow cartogram, 268
Orthographic projection, 147-48

Pace scale, 29
Pakistan, 82, 85
Panchnad Nadi, 60
Pantograph, 40
Parallax Bar, 316
Pencker, K., 97
Physical maps, 35
Physiographic information, 57
Plain scale, 6
Plane table survey, 168
Pocket stereoscope, 315
Polyconic projection, 138
Population pyramid, 267
Precipitation, 67, 78
Presmatic compass survey, 177
close traverse, 180
open traverse, 179
Pressure, 80, 81
altimeter, 74
distribution of, 82-83
gradient, 83
relation, 327
Profiles, 115
Psychrometer, 69

Quartiles, 210-28
deviation, 223-26

Railway survey, 160
Rainfall, 80
Rain gauge, 67
Rawa Nandi, 59, 60
Relief maps, 36
Revolution scale, 22
Ria coast, 115
Ridge, 105
Ring diagram, 276
Rocks, 293
extrusive, 294
intrusive, 293
igneous, 293, 297
kinds of, 297-98
metamorphic, 296
type of, 301-302
sandy, 295
type of, 299-300
Robinson, A.H., 99

Saddle, 105
Sand-dunes, 115
Scale,
comparative, 11
construction, 6-7
geometrical method, 9
methematical method, 9
definition, 1
diogonal, 23
method, 1
conversation, 4
graphical, 2,5
plain, 6
representation fraction, 2
simple statement, 1
vernier, 28
Scanning stereoscope, 316
Sea condition, 80-81, 84
Second World War, 11
Sextant, 188-92
Siwalik ranges, 59
Sketch master, 316
Slops, 93
kinds, 93-94
methods of, 95
Raisz and Henry, 99
Robinson, 99
Smith, 98
Wentworth, 98
relief representation, 97, 98, 99
Smith, G. H., 98
Soil maps, 36
Sonah Nadi, 60
Spherical diagram, 280
Spur, 105
Sri Lanka, 82
Standard Deviation, 232-38
Star diagram, 268
Statistical data,
distribution map, 285-87
one dimensional diagram, 253-270
representation of, 253-83
three dimensional diagram, 280-81
Statistical method, 199-252
central tendency measures of, 200-235
Stereographic projection, 144
Stevensions' screen, 62
Sunshine, 72
Survey of India maps, 52
Carte international Du monde series, 52
India and adjacent countries series, 52

southern Asia series, 52
Surveying, 159-198
Abney's level, 194
classification, 159-60
chain surveying, 160-62
definition, 159
Indian climometer, 192
levelling, 184-85
methods of, 172-74, 175-77
intersection, 174
radiant, 173
re-section, 175
tracing paper, 177
traversing, 174
plane table, 168
presmatic compass survey, 197
sextant, 188
tapes, 162
theodolite, 197

Taylor, Graffic, 260
Tele thermometer, 65
Temperature relation, 327
Theodolite, 197-98
Thermograph, 65
Three Dimensional diagrams, 280-81
Time and distance scale, 16
Topographical map, 1
survey, 159
Topo-sheet, 60
arrangement, 53
interpretation, 51-52
studies, 59-60
Traffic flow cartogram, 268
Tributaries, 60

U-shaped valley, 112

V-shaped valley, 112
Vegetation, 60
Vernier scale, 28
construction, 30
Velconic cones, 108

Walder E,Finstar, 97
Wall paper, 35
Water fall, 112
Weather elements, 80
Weather maps, 36, 62
daily report, 81
element, 38
instruments, 62
Weight and Measurement Unit, 325-26
Wentworth, C.K., 98
Wet bulb and dry bulb thermometer, 64
Wheel diagram, 276
Winds, 80, 83

Yard scale, 22

Zenittal equidistant projection, 141, 148-50